Compendium of Plant Genomes

Series Editor

Chittaranjan Kole, Raja Ramanna Fellow, Government of India, ICAR-National Research Center on Plant Biotechnology, Pusa, New Delhi, India

AF823746

Whole-genome sequencing is at the cutting edge of life sciences in the new millennium. Since the first genome sequencing of the model plant Arabidopsis thaliana in 2000, whole genomes of about 100 plant species have been sequenced and genome sequences of several other plants are in the pipeline. Research publications on these genome initiatives are scattered on dedicated web sites and in journals with all too brief descriptions. The individual volumes elucidate the background history of the national and international genome initiatives; public and private partners involved; strategies and genomic resources and tools utilized; enumeration on the sequences and their assembly; repetitive sequences; gene annotation and genome duplication. In addition, synteny with other sequences, comparison of gene families and most importantly potential of the genome sequence information for gene pool characterization and genetic improvement of crop plants are described.

Interested in editing a volume on a crop or model plant? Please contact Prof. C. Kole, Series Editor, at ckoleorg@gmail.com

More information about this series at http://www.springer.com/series/11805

Maizura Ithnin • Ahmad Kushairi
Editors

The Oil Palm Genome

Springer

Editors
Maizura Ithnin
Advanced Biotechnology
and Breeding Centre
Malaysian Palm Oil Board (MPOB)
Kajang, Malaysia

Ahmad Kushairi
Malaysian Palm Oil Board (MPOB)
Kajang, Malaysia

ISSN 2199-4781 ISSN 2199-479X (electronic)
Compendium of Plant Genomes
ISBN 978-3-030-22548-3 ISBN 978-3-030-22549-0 (eBook)
https://doi.org/10.1007/978-3-030-22549-0

© Springer Nature Switzerland AG 2020
This work is subject to copyright. All rights are reserved by the Publisher, whether the whole or part of the material is concerned, specifically the rights of translation, reprinting, reuse of illustrations, recitation, broadcasting, reproduction on microfilms or in any other physical way, and transmission or information storage and retrieval, electronic adaptation, computer software, or by similar or dissimilar methodology now known or hereafter developed.
The use of general descriptive names, registered names, trademarks, service marks, etc. in this publication does not imply, even in the absence of a specific statement, that such names are exempt from the relevant protective laws and regulations and therefore free for general use.
The publisher, the authors and the editors are safe to assume that the advice and information in this book are believed to be true and accurate at the date of publication. Neither the publisher nor the authors or the editors give a warranty, expressed or implied, with respect to the material contained herein or for any errors or omissions that may have been made. The publisher remains neutral with regard to jurisdictional claims in published maps and institutional affiliations.

This Springer imprint is published by the registered company Springer Nature Switzerland AG
The registered company address is: Gewerbestrasse 11, 6330 Cham, Switzerland

This book series is dedicated to my wife Phullara and our children Sourav and Devleena

Chittaranjan Kole

Preface to the Series

Genome sequencing has emerged as the leading discipline in the plant sciences coinciding with the start of the new century. For much of the twentieth century, plant geneticists were only successful in delineating putative chromosomal location, function, and changes in genes indirectly through the use of a number of "markers" physically linked to them. These included visible or morphological, cytological, protein, and molecular or DNA markers. Among them, the first DNA marker, the RFLPs, introduced a revolutionary change in plant genetics and breeding in the mid-1980s, mainly because of their infinite number and thus potential to cover maximum chromosomal regions, phenotypic neutrality, absence of epistasis, and codominant nature. An array of other hybridization-based markers, PCR-based markers, and markers based on both, facilitated construction of genetic linkage maps, mapping of genes controlling simply inherited traits, and even gene clusters (QTLs) controlling polygenic traits in a large number of model and crop plants. During this period, a number of new mapping populations beyond F2 were utilized and a number of computer programmes were developed for map construction, mapping of genes, and mapping of polygenic clusters or QTLs. Molecular markers were also used in the studies of evolution and phylogenetic relationship, genetic diversity, DNA fingerprinting, and map-based cloning. Markers tightly linked to the genes were used in crop improvement employing the so-called marker-assisted selection. These strategies of molecular genetic mapping and molecular breeding made a spectacular impact during the last one and a half decades of the twentieth century. But still they remained "indirect" approaches for elucidation and utilization of plant genomes since much of the chromosomes remained unknown and the complete chemical depiction of them was yet to be unravelled.

Physical mapping of genomes was the obvious consequence that facilitated the development of the "genomic resources" including BAC and YAC libraries to develop physical maps in some plant genomes. Subsequently, integrated genetic–physical maps were also developed in many plants. This led to the concept of structural genomics. Later on, emphasis was laid on EST and transcriptome analysis to decipher the function of the active gene sequences leading to another concept defined as functional genomics. The advent of techniques of bacteriophage gene and DNA sequencing in the 1970s was extended to facilitate sequencing of these genomic resources in the last decade of the twentieth century.

As expected, sequencing of chromosomal regions would have led to too much data to store, characterize, and utilize with the-then available computer software could handle. But the development of information technology made the life of biologists easier by leading to a swift and sweet marriage of biology and informatics, and a new subject was born—bioinformatics.

Thus, the evolution of the concepts, strategies, and tools of sequencing and bioinformatics reinforced the subject of genomics—structural and functional. Today, genome sequencing has travelled much beyond biology and involves biophysics, biochemistry, and bioinformatics!

Thanks to the efforts of both public and private agencies, genome sequencing strategies are evolving very fast, leading to cheaper, quicker, and automated techniques right from clone-by-clone and whole-genome shotgun approaches to a succession of second-generation sequencing methods. The development of software of different generations facilitated this genome sequencing. At the same time, newer concepts and strategies were emerging to handle sequencing of the complex genomes, particularly the polyploids.

It became a reality to chemically—and so directly—define plant genomes, popularly called whole-genome sequencing or simply genome sequencing.

The history of plant genome sequencing will always cite the sequencing of the genome of the model plant Arabidopsis thaliana in 2000 that was followed by sequencing the genome of the crop and model plant rice in 2002. Since then, the number of sequenced genomes of higher plants has been increasing exponentially, mainly due to the development of cheaper and quicker genomic techniques and, most importantly, the development of collaborative platforms such as national and international consortia involving partners from public and/or private agencies.

As I write this Preface for the first volume of the new series *Compendium of Plant Genomes*, a net search tells me that complete or nearly complete whole-genome sequencing of 45 crop plants, eight crop and model plants, eight model plants, 15 crop progenitors and relatives, and three basal plants is accomplished, the majority of which are in the public domain. This means that we nowadays know many of our model and crop plants chemically, i.e. directly, and we may depict them and utilize them precisely better than ever. Genome sequencing has covered all groups of crop plants. Hence, information on the precise depiction of plant genomes and the scope of their utilization are growing rapidly every day. However, the information is scattered in research articles and review papers in journals and dedicated Web pages of the consortia and databases. There is no compilation of plant genomes and the opportunity of using the information in sequence-assisted breeding or further genomic studies. This is the underlying rationale for starting this book series, with each volume dedicated to a particular plant.

Plant genome science has emerged as an important subject in academia, and the present compendium of plant genomes will be highly useful to both students and teaching faculties. Most importantly, research scientists involved in genomics research will have access to systematic deliberations on the plant genomes of their interest. Elucidation of plant genomes is of interest not only for the geneticists and breeders, but also for practitioners of an array of plant science disciplines, such as taxonomy, evolution, cytology, physiology, pathology, entomology, nematology, crop production, biochemistry, and obviously bioinformatics. It must be mentioned that information regarding each plant genome is ever-growing. The contents of the volumes of this compendium are, therefore, focusing on the basic aspects of the genomes and their utility. They include information on the academic and/or economic importance of the plants, description of their genomes from a molecular genetic and cytogenetic point of view, and the genomic resources developed. Detailed deliberations focus on the background history of the national and international genome initiatives, public and private partners involved, strategies and genomic resources and tools utilized, enumeration on the sequences and their assembly, repetitive sequences, gene annotation, and genome duplication. In addition, synteny with other sequences, comparison of gene families, and, most importantly, the potential of the genome sequence information for gene pool characterization through genotyping by sequencing (GBS) and genetic improvement of crop plants have been described. As expected, there is a lot of variation of these topics in the volumes based on the information available on the crop, model, or reference plants.

I must confess that as the series editor, it has been a daunting task for me to work on such a huge and broad knowledge base that spans so many diverse plant species. However, pioneering scientists with lifetime experience and expertise on the particular crops did excellent jobs editing the respective volumes. I myself have been a small science worker on plant genomes since the mid-1980s, and that provided me the opportunity to personally know several stalwarts of plant genomics from all over the globe. Most, if not all, of the volume editors are my long-time friends and colleagues. It has been highly comfortable and enriching for me to work with them on this book series. To be honest, while working on this series I have been and will remain a student first, a science worker second, and a series editor last. And I must express my gratitude to the volume editors and the chapter authors for providing me the opportunity to work with them on this compendium.

I also wish to mention here my thanks and gratitude to the Springer staff, particularly Dr. Christina Eckey and Dr. Jutta Lindenborn for the earlier set of volumes and presently Ing. Zuzana Bernhart for all their timely help and support.

I always had to set aside additional hours to edit books beside my professional and personal commitments—hours I could and should have given to my wife, Phullara, and kids, Sourav and Devleena. I must mention that they not only allowed me the freedom to take away those hours from them but also offered their support in the editing job itself. I am really not sure whether my dedication of this compendium to them will suffice to do justice to their sacrifices for the interest of science and the science community.

New Delhi, India Chittaranjan Kole

Preface

Genomics is a combined discipline of classical genetics and computational science. Genomics aims at understanding the structure, function, evolution, genetic mapping, and genome editing of an organism using DNA sequence data. This book showcases significant breakthroughs and updates in oil palm genomic research and related fields. Meant for a wide spectrum of readers, a chapter on the history and economic importance of the crop is included. With oil palm genome sequence, genes or markers associated with agronomic traits of interest were identified while new tools were developed. Certain tools were applied to estimate genetic diversity of the oil palm. Such information is crucial to breeders in designing appropriate breeding schemes to enrich the narrow genetic base of current planting materials for sustainable development of the industry. The genetic control of economically important phenotypes, such as fruit forms, fruit colours, and tissue culture-related abnormalities, is among the principle outcomes from genomic research. Subsequent chapters highlight the introgression of wild germplasm with current genetic materials and applying modified reciprocal recurrent selection scheme in breeding programmes and production of improved planting materials. Genetic improvements were further enhanced by means of molecular cytogenetics tools and marker-assisted selection, developed from genome-wide association studies (GWAS) and genomic selection (GS). This method is to select palms carrying specific chromosomes and favourable QTLs more precisely for breeding programmes, leading to the development of elite planting materials. Such high-yielding materials with niche characteristics are tissue cultured to obtain a large number of uniformed planting materials. An array of innovations were developed including fine-tuning the propagation protocols to more user-friendly and efficient with fast result. Genomic-derived quality control tool was developed and applied across tissue culture materials to minimize yield-affecting somaclonal variations. Complementing the breeding techniques, genetic engineering is used to diversify palm oil applications vis-a-vis higher value-added products. Researchers continuously working towards optimizing genetic transformation systems of the oil palm and challenges faced during the process deliberated. The final chapter presents the state-of-the-art post-genomics tools such as transcriptomics, proteomics, and metabolomics which are embraced as phenotyping tools to elucidate the mechanisms in fruit ripening and fatty acid synthesis, among others. On the account of the indispensable need to unravel diseases in oil palm, post-genomic tools are exploited to advance knowledge in plant–pathogen interaction for novel biomarker discoveries. The chapters in this book were contributed by experts in their respective fields of research. This book provides a comprehensive reference material in genomics research for both oil palm and non-oil palm scientists.

Kajang, Malaysia

Maizura Ithnin
Ahmad Kushairi

Contents

1 **The History and Economic Importance of the Oil Palm** 1
Sean Mayes

2 **Wild and Advanced Resources of *Elaeis guineensis* and *Elaeis oleifera*** 9
Mohd Din Amiruddin, Nor Azwani Abu Bakar, Fadila Ahmad Malike, Suzana Mustaffa, Norziha Abdullah, Marhalil Marjuni, Mohd Mustakim Mohamad, Mohd Yazid Hassan, Ahmad Kushairi, and Rajanaidu Nookiah

3 **Breeding and Improvement of the Oil Palm Interspecific Hybrids via Molecular Cytogenetics** 25
M. Madon, H. M. N. Nordiana, Z. Muhammad Azwan, and N. M. Zaki

4 **Genetic Improvement of Oil Palm Through Recurrent Selection** 35
Mohd Rafii Yusop, Jalani Sukaimi, Mohd Din Amiruddin, Momodu Jalloh, Senesie Swaray, Oladosu Yusuff, and Samuel Chibuike Chukwu

5 **Oil Palm Tissue Culture: Fast Tracking Elite Commercial Lines** 47
Zamzuri Ishak, Ahmad Tarmizi Hashim, Samsul Kamal Rosli, Dalilah Abu Bakar, Siew Eng Ooi, Naqiuddin Mohd, and Meilina Ong-Abdullah

6 **Oil Palm Transgenic Research: Challenges, Update, and Future Outlook** 69
Omar Abdul Rasid, Subhi Siti Masura, Ayub Nor Hanin, Dayang Izawati Abang Masli, Bahariah Bohari, Mat Yunus Abdul Masani, and Ghulam Kadir Ahmad Parveez

7 **Oil Palm Genome: Strategies and Applications** 83
Eng-Ti Leslie Low, Norazah Azizi, Mohd Amin Ab Halim, Nik Shazana Nik Mohd Sanusi, Kuang-Lim Chan, Nadzirah Amiruddin, Nagappan Jayanthi, Meilina Ong-Abdullah, Rajinder Singh, Ravigadevi Sambanthamurthi, Mohamad Arif Abdul Manaf, and Ahmad Kushairi

8 Molecular Evolution and Genetic Diversity of Oil Palm Based on Sequencing and Analysis with Molecular Markers . . . 117
Parisa Azizi, Mohamed Musa Hanafi, Mohamad Osman, Mohd Rafii Yusop, Maizura Ithnin, Sima Taheri, and Mahbod Sahebi

9 Genetic Dissecting Complex Traits via Conventional QTL Analysis and Association Mapping . . . 131
Chee-Keng Teh, Nurshazwani Amalina Sudirman, Fairuz Farhana Mohd Rodzik, Ai-Ling Ong, Qi-Bin Kwong, and David Ross Appleton

10 Omics—A Potential Tool for Oil Palm Improvement and Productivity . . . 141
Umi Salamah Ramli, Abrizah Othman, Noor Idayu Mhd Tahir, Benjamin Yii Chung Lau, Syahanim Shahwan, Hasliza Hassan, Zain Nurazah, Shahirah Balqis Dzulkafli, Nurul Liyana Rozali, Nur Ain Ishak, Ravigadevi Sambanthamurthi, Ghulam Kadir Ahmad Parveez, and Ahmad Kushairi

The History and Economic Importance of the Oil Palm

1

Sean Mayes

Abstract

Oil palm is a remarkable crop which in 2017 produced 34% of the world's vegetable oil from 0.36% of the world's agricultural land. The fact that it is a perennial 'tree' crop growing in the humid tropics under high light intensity accounts for the tenfold advantage in oil yield it has over many annual temperate oil crops, per hectare, per year. However, slow breeding cycles in a manually intensive industry are beginning to erode this advantage, and the challenges of climate change have yet to be addressed. The history, economic importance and how research has been applied to improve the productivity and value of the oil palm are reviewed, along with some of the future challenges which research needs to address.

1.1 Introduction

The oil palm is a recent crop success story, particularly in Southeast Asia. While it was widely known and traded during the European industrial revolution, it has only really become a major crop in the last century. This is largely due to the establishment of a wide range of food and nonfood uses and of organised plantations which now produce the majority of the world's traded palm oil and vegetable oil as well. The fundamental basis for this success has been productivity, with oil yields now more than fourfold higher than at plantation establishment in the 1920s (Corley and Lee 1992; Henson 2012—Table 1c). With the current average yields in Malaysia, for example, running at between 3.5 and 4 tonnes oil per hectare per year, this is also around tenfold the yield obtained by some annual oil crops. Moreover, global production of vegetable oil has increased from 90.5 million metric tonnes in 2000/01 to 195.1 million metric tonnes in 2017/18, an increase which has been driven by rising demand and led to palm oil being the most traded global oil (Statistica; 27-08-2018).

While the recent past of oil palm cultivation has been a success story, the future is more uncertain. The yield gap between average yield actually obtained and potential yield is stubbornly static in many countries with the best breeding trials achieving 10–12 tonnes of oil per hectare per year (Henson 2012), and with plantations facing significant palm disease and labour shortage (and cost) threats, oil palm as an industry will need to change significantly in the future if it is to remain competitive and relevant. This is further exacerbated by the poor press particularly relating to deforestation (Vijay et al. 2016) that the oil palm industry receives in some

S. Mayes (✉)
University of Nottingham, Sutton Bonington Campus, Loughborough, Leicestershire LE12 5RD, UK
e-mail: sean.mayes@nottingham.ac.uk

© Springer Nature Switzerland AG 2020
M. Ithnin and A. Kushairi (eds.), *The Oil Palm Genome*, Compendium of Plant Genomes, https://doi.org/10.1007/978-3-030-22549-0_1

of its biggest markets, such as Europe, and the current predictions for climate change and their potential impacts (Paterson et al. 2015).

Corley and Tinker (2015) cover just about everything to do with oil palm and the industry, while Henson (2012) also provides a detailed review of the history of oil palm. Henderson and Obsorne (2000) provide an interesting take on some of the drivers for the industry developing.

The current chapter provides a brief overview of the history and importance of this crop and looks at some of the research history, while also identifying some of the broad challenges for the future.

1.2 Taxonomy, Biology and Distribution

Botanically, 'African oil palm' is classified in order, Arecales; family, Arecaceae; subfamily, Arecoideae; tribe, Cocoseae and in the genus 'Elaeis' with the specific name reflecting the presumed origin in African Guinea, hence *Elaeis guineensis*. It is a monoecious palm with a single growing point, which produces pinnate leaves and bunches of fruit composed of drupes with an endocarp, shell and exocarp. Palm oil comes from the exocarp and kernel oil from the endocarp. For mature palms, bunches can be composed of thousands of fruits and weigh around 50 kg. Individual palms cycle between the production of male and female inflorescences, naturally promoting outcrossing, although some bunches may have small numbers of hermaphroditic flowers (Corley and Tinker 2015).

The wild/semi-wild palm belt in Africa stretches from roughly +10° N to 10° S (although populations are also found in Madagascar, some 20° S, although appear to be relatively genetically distinct (Corley and Tinker 2015—Fig. 1; Bakoume et al. 2015). While *E. guineensis* has been found growing 'wild' in South America, pollen and fossil evidence suggest an African origin (Zeven 1964), with potential transfer of African oil palm to South America in historical time, where it was only picked up industrially in recent years.

A related palm, *E. oleifera,* is also within the genus of Elaeis and is the 'South American oil palm'. The consensus is that the two species split with the breakup of an ancient landmass around 60 million years ago, leading to a speciation by geographical separation without genetic barriers. While hybrids are possible, they can have limited fertility, requiring backcrossing. Production based on F_1 hybrids requires hand-pollination to ensure bunch set. Both species have 16 chromosomes ($2n = 32$), and cytogenetic studies have shown three classes of chromosomes according to length, with similarity between the two species (Maria et al. 1995, 1998; Castilho et al. 2000) although Genomic In Situ Hybridisation (GISH) can be used to distinguish the species (Cheah et al. 2000; Zaki et al. 2017). The recent publication of the genome sequence for *E. guineensis* and comparisons to *E. oleifera* show extensive genomic differences, with substantial variation in the types and quantities of repeat elements between species (Price et al. 2002, 2003; Singh et al. 2013a). Divergence tests suggested a separation time consistent with the geographical separation theory (51 MYR) (Singh et al. 2013a) arguing for a very ancient divergence point, with subsequent genomic evolution.

Genetic diversity analysis from the collections made by Palm Oil Research Institute of Malaysia (PORIM; now Malaysian Palm Oil Board—MPOB) shows the highest level of variation for *E. guineensis* within a Nigerian collection, when compared with 11 other African countries in the palm belt (Maizura et al. 2006) suggesting this region as a possible centre of origin or diversity.

1.3 History and Economic Importance

The first recorded sample of what has been tentatively identified as palm oil was rin from funeral goods interned with a Pharaoh of the Early Dynastic period, around 5000 BCE (Raymond 1961). This suggests that palm oil was recognised to be of value and that human exploitation of palm oil was already underway at that point. Excavations in Ghana have identified preserved samples of palm oil from around 4500 BCE indicating that local exploitation in West Africa was underway at a similar period to the Abydos sample in Egypt. The Abydos sample also argues for the 'trade'or at least movement of oil palm, given that oil palm is unlikely to have ever grown in Egypt. It has been suggested that Arab traders were responsible for taking it to Egypt (Obahiagbon 2012).

The first botanical description was produced by Jacquin (1763) after whom the species is named, with more detailed description by Gaertner (1788) with a recognition of the ability of the palm to produce inflorescences of different sexes. Likely written descriptions by Portuguese explorers date from 1435 to 1460 (Crone 1937). Palm oil was used as part of the food for slaves during the slave trade on the long journey to the New World, but it was really developments in Europe which secured the future of the African oil palm trade. The Industrial Revolution led to the development of saponification and created a major demand for oil palm for lubricants, detergents (e.g. Palmolive, being a well-known soap brand)(Henderson and Obsorne 2000) and other industrial products, and this demand saw trade (still as a cottage industry) increase significantly. Eventually, the trade in oil palm was promoted as a way to suppress the continuing trade in slaves, which was banned in Britain and the British Empire in 1807 through the efforts of William Wilberforce and the abolitionists, but continued clandestinely for many years (Stilliard 1938), with the trade in palm oil seen as a legitimate replacement for shipping slaves from Africa.

The quality of the initial palm oil was poor—it was still essentially a cottage industry, and oil and kernels would be prepared by the villages along the coast or imported from further inland. It was only really with the establishments of the first forts that a more coherent approach to oil palm trade begins, and eventually, this led to the establishment of the first plantations in colonial Africa. In the 1880s, around 75% of Ghana's international trade was in palm oil (http://mofa.gov.gh/site/?page_id=8819). One of the most notable early plantations was set up on a Belgian concession in the Congo (Democratic Republic of Congo at Binga) in 1935 by Sir William Lever, who founded the British part of Unilever (Henderson and Obsorne 2000).

Since these initial stages of industrial development, considerable progress has been made, although the focus of the industry shifted after World War II towards Southeast Asia, with significant expansion in the 1960s and 1970s, where plantations established in Malaysia and Indonesia were highly productive and now produce 85% of palm oil in the world.

A concern of the current oil palm industry is a relatively restriction origin for some of the most important origins in the current commercial use. The final commercial palm is a hybrid between thick-shelled *dura* palms and shell-less *pisifera* pollen parents. This single gene trait has had more influence on palm breeding than any other (Beirnaert and Vanderweyen 1941; Singh et al. 2013b). According to pedigree, the main Deli *dura* population (seed parent) is derived from four palms introduced into the Bogor Botanic Garden in Indonesia in 1848. Offspring of these were grown in the Economic Garden and were used by many plantations for selection, although selection criteria adopted were often different, leading to a number of Breeding Populations of Restricted Origin (BPRO) (Rosenquist 1986). On the pollen side, the offspring of a palm called 'Djongo'—the 'best'—was imported from Congo and eventually came to represent the AVROS *pisifera* origin with 75% of the genetic material derived from a single palm. This is now the most important *pisifera* source (Corley and Tinker 2015). However, as oil palm is naturally outcrossing, going through cycles of male and female inflorescence production, it seems very likely that early attempts at controlled pollination involved relatively high levels of pollen contamination. The introduction of the pollinating weevil, *Elaeidobius kamerunicus* into Southeast Asia in 1981 (Syed 1982) led to high levels of *dura* contamination in controlled commercial seed production, so the original African sources of germplasm, where the pollinating weevils were endemic, are likely to represent a range of material from different pollen sources, even if the maternal palm was correctly selected.

Despite a technically very narrow pedigree, the performance of the introduced thick-shelled *dura* fruit type was sufficiently good (or the tested African thin-shelled *tenera* sufficiently bad) that Malaysia and Indonesia only moved reluctantly from planting thick-shelled *dura* as commercial material to thin-shelled *tenera* in the 1960s. This shift increases oil yields through reducing shell thickness and increasing mesocarp oil by 30%. It certainly is also the case that the environment in Southeast Asia is more favourable than much of Africa, being humid with high rainfall, stable temperatures, good light and, perhaps most importantly, little in the way of seasonal variation, so that palms can produce all year round. Today, Southeast Asia accounts for around 90 per cent of world production and 61% of world trade. In 2017, 67.92 million tonnes of palm oil were produced, together with 7.25 million tonnes of palm kernel oil, compared with 53.94 million tonnes of soybean oil, the second-largest oil producer. Oil palm is planted on 19.04 million hectares of world agricultural land (0.36%). Oil palm alone accounts for 34% of world production (Kushairi et al. 2018).

Oil palm is also unusual in that it produces two oils: mesocarp oil from the fruit pericarp and kernel oil from the kernels. These are extracted separately and have different compositions, with 'crude palm oil': CPO—mesocarp oil—being roughly 50% palmitic, 40% oleic and 10% other unsaturated oils. This makes it more solid at (nontropical) room temperatures, and it is often fractionated into oleic and stearic fractions. The kernel oil is formed of saturated short-chain fatty acids, which provide a composition quite similar to coconut oil, and it is often used in ice creams and coffee whiteners. Because of the desire to separately extract the two oils, the processing is slightly more complicated than for temperate oilseed crops. In addition to the oil itself, CPO is high in carotenoids and tocopherols and tocotrienols. The first can be used for Vitamin A production and the latter ones for Vitamin E production. The crushed seed cake can be fed to animals (but is a relatively poor feed), and palm oil can be converted to biodiesel, while there is also some use of the palm trunk for construction materials (Soh et al. 2017a).

1.4 The Challenges

Breeding in any tree crop is often a long-term process, and oil palm is no exception. The understanding that the control of the fruit shell thickness was under control of a single gene means that almost all commercially planted material today is of *tenera* hybrids. In practice, this has led to variations on Recurrent Reciprocal Selection (RRS) and Family and Individual Selection (FIS) schemes, or their combinations (Soh et al. 2017b). Because of the requirement for hybrids and the difficulty predicting breeding values in the *pisifera* line (many origins are female sterile due to bunch abortion), the actual breeding cycle for oil palm is long (between 12 and 19 years). Despite the long cycles, an experiment evaluating different rounds of selection of material suggests that in four generations, the yield had quadrupled, compared to unimproved material. Roughly half of this was attributed to genetic improvement and half to improvements in management and agronomy (Corley and Lee 1992).

The main factors influencing response to breeding selection are the heritability of the trait and the selection intensity imposed. 'Oil yield per palm' has a lower heritability than 'oil-to-mesocarp ratio', and greater response to selection would be expected by focusing on more heritable components of the desired oil yield trait. This led to the idea of bunch analysis (Blaak 1965) which breaks down oil yield into the component traits which underlie it. This approach to assessing palm value in a breeding programme has been largely unchanged since it was originally adopted. In a breeding trial, it is possible to measure fresh fruit production from individual palms, but it is not possible to bunch analyse all of the bunches from a palm. This leads to a sampling approach, where the value of a palm is only judged when a minimum of three bunches have been analysed (and preferably over five) and often only as a part of assessing the quality of the parental palms of a family. While this appears potentially limiting in terms of selection accuracy, it clearly works in practice, with yield potential continuing to advance. Finding a less labour-intensive or more accurate approach could be an important area to improve the efficiency (and potentially accuracy) of future selection (Corley 2018a, b).

The presence of only a single vegetative meristem at the apex of the palm spurred the development of tissue culture approaches to allow the clonal multiplication of this heterozygous palm. In practice, the methods developed by a number of organisations in the 1960s and 70s focused around somatic embryogenesis with the intention to mass produce the elite palms clonally (Jones 1974; Rabechault and Martin 1976). Regeneration of palms is relatively slow (around two years in many cases), and there is a strong genotype dependence for embryogenesis. However, after a series of hormonally induced stages, the process works and can be highly productive. It faces two main problems: accuracy of predicting the performance of clones and, more critically, the development of abnormality for flowering in clones. The latter was first reported in Unilever material when it was observed that continuous culturing of material in tissue culture led to progressively more palms which produced 'mantled fruits' where the rudimentary androecium in female inflorescences develops into supplementary carpels (Corley et al. 1986). The problem that this leads to is poor fruit set and bunch abortion. While it has been shown that there is clearly a general decrease in methylation during tissue culture which can persist after regeneration, the oil palm at the sequence level is remarkably stable. The locus responsible for the mantling phenotype has recently been identified. This was predicted to be a homeotic floral gene from soon after the phenomenon was first reported, but has only recently been elucidated as changes to methylation patterns (CHG) in a retrotransposon element in an intron of the *Deficiens* gene of oil palm (Ong-Abdullah et al. 2015, 2017). Had this locus not been sensitive to tissue culture, the current planting material and breeding schemes in oil palm would have been very different today. However, the development of an accurate diagnostic test could allow the potential of clonal propagation to be finally realised, and developments in the field of molecular markers and (in the future) phenotyping could address the relatively poor correlation between parental palm performance (Ortet) and clonal offspring performance (Ramet).

1.5 Overseeing the Challenges

A major step forward was the public release of the genome sequence of a *pisifera* palm by the Malaysian Palm Oil Board (MPOB) (Singh et al. 2013a) quickly followed by reports of the cloning of the genes for shell thickness (Singh et al. 2013b), *virescens* (Singh et al. 2014) and identification of the gene responsible for the mantled (abnormal) flowering form which has dogged the application of tissue culture for over 40 years (Ong-Abdullah et al. 2015).

The identification of these genes potentially allows the development of perfect markers, which test the actual site of mutation responsible for the phenotype. Commercial diagnostic test kits have been developed for shell and fruit colour assay, while it is under development for clonal abnormality. However, the accuracy of such kits depends on the specific allelic variant present in the different sources of germplasm, and it may be a number of years before the kits are 100% accurate and account for all variants within oil palm germplasm.

Alongside the identification of these major genes which influence oil palm tissue culture and breeding, other groups (and notably Sime Darby Plantations in Malaysia) have focused on the development of high-density marker systems and the routine application of marker information integrated with the breeding programme. This has allowed the integrated use of 'Omic' technology in gene discovery (Teh et al. 2017a), but perhaps more importantly the development of association genetic analysis and the testing of genomic selection methods. Genomic selection and variants of what might be termed 'informed genomic selection' where GWAS results are used to inform the models created, address the biggest hurdle for oil palm breeding- selection cycle length (Wong and Bernado 2008; Kwong et al.

2016, 2017; Cros et al. 2015a, b; Cros 2017a, b; Teh et al. 2017b).

However, the biggest hurdle overall facing the industry at the moment in Southeast Asia is cost of labour for harvesting, which is highly manual at the moment. The leaf subtending the bunch needs to be removed using a curved blade on an aluminium pole, before the bunch stalk can be cut. For older plantations, the bunch can fall several metres, and many fruits detach on landing. This necessitates the collection of loose fruits from the ground as well as the impact activating the lipases which begin to degrade the quality of the oil in the mesocarp.

Further domestication is clearly a major target for future breeding and application of biotechnology with the design of new ideotypes (e.g. Corley 2017), including a redesign of the bunch structure itself. This could be an important step forward to reduce the dependency on manual harvesting and to modernise plantation practices. At the same time as there is a need to reduce inputs for increased plantation sustainability (both ecological and economic), the effects of climate change are beginning to be felt, particularly in the form of periods of relative drought in Malaysia and these are predicted to impact further (Paterson et al. 2015; Soh et al. 2017c). This makes breeding for a degree of drought tolerance more important for some regions or for the more extensive use of water management and, perhaps in the longer term, irrigation, which will add to costs. The recent drought episodes in Malaysia have seen a direct fall in fruit production as well as potentially storing up longer term problems (stress can lead to a decreased sex ratio, with more male influorences being formed and emerging over two years later).

1.6 Conclusion

While oil palm faces a potentially difficult future given the number of pressures on it, ultimately, it is likely to survive as an industry. The hard fact is that oil palm produces 34% of the world's vegetable oil on 0.36% of the agricultural land. There is, at this point in time, no crop which can currently replace it for productivity and which is economically viable. With the rapid advances in genetic technology (and particularly the next-generation sequencing and 'Omics'), the tools are available to meet these challenges from a breeding perspective. Supplementing these with specific applications of tissue culture and gene editing provides a powerful tool kit to develop future oil palm. There is a clear need for better phenotyping methods and the development of high throughput approaches appropriate for oil palm (so called phenomics), but the foundations for an ecologically and economically sustainable industry do exist and through hard work and focused application of science, can be achieved (Soh et al. 2017c). While improvements in oil palm are required, looking again at the economics of the plantation and land use may allow multiple crops to be grown, improving plantation economic sustainability and also agricultural biodiversity.

References

Bakoume C, Wickneswari R, Siju S, Rajanaidu N, Kushairi A, Billotte N (2015) Genetic diversity of the world's largest oil palm (*E. guineensis* Jacq.) field genebank accessions using microsatellite markers. Genet Resour Crop Evolut 62:349–360

Beirnaert A, Vanderweyen R (1941) Contribution a l'etudegenetique et biometrique des varieties d'*Elaeis-guineensis* Jacquin. Publ Inst Nat Etude Agron Congo Belge Ser Sci 27:1–101

Blaak G (1965) Breeding and inheritance in oil palm (*Elaeisguineensis* Jacq) Part III: yield selection and inheritance. J West Afr Inst Oil Palm Res 4:262–283

Castilho A, Vershinin A, Heslop-Harrison JS (2000) Repetitive DNA and the chromosomes in the genome of oil palm (*Elaeisguineensis*). Ann Bot 85:837–844

Cheah SC, Madon M, Singh R (2000) Oil palm genomics. In: Basiron Y, Jalani BS, Chan KW (eds) Advances in oil palm research, vol I. Malaysian Palm Oil Board, Kuala Lumpur, pp 332–370

Corley RHV (2017) 13.9: Summing up: an ideotype for yield and sustainability. In: Soh AC, Mayes S, JAR Roberts (eds) Oil palm breeding: genetics and genomics. CRC Press, Taylor and Francis Group, New York, pp 397–405

Corley RHV (2018a) Studies of bunch analysis 1—variation within and between palms. J Oil Palm Res 30 (2):196–205

Corley RHV (2018b) Studies of bunch analysis 2—bunch sampling to estimate oil yield. J Oil Palm Res 30 (2):206–218

Corley RHV, Lee CH (1992) The physiological basis for genetic improvement of oil palm in Malaysia. Euphytica 60:179–184

Corley RHV, Tinker PB (2015) The oil palm, 5th edn. World Agricultural Series, Wiley Blackwell; ISBN-10:1405189398

Corley RHV, Lee CH, Law IH, Wong CY (1986) Abnormal flower development in oil palm clones. Planter 62:233–240

Crone GR (1937) The voyages of Cadamosto and other documents on Western Africa in the second half of the fifteenth century. Hakluyt Soc. Series II, 80

Cros D (2017a) Appendix 6D1; Principles of genomic selection; breeding plans and selection methods. In: Soh AC, Mayes S, Roberts JAR (eds) Oil palm breeding: genetics and genomics. CRC Press, Taylor and Francis Group, New York, pp 156–160

Cros D (2017b) Genomic Selection for oil palm. In: Soh AC, Mayes S, Roberts JAR (eds) Oil palm breeding: genetics and genomics. CRC Press, Taylor and Francis Group, New York, pp 263–269

Cros D, Denis M, Bouvet JM, Sanchez L (2015a) Long-term genomic selection for heterosis without dominance in multiplicative traits: case study of bunch production in oil palm. BMC Genom 16:651

Cros D, Deni M, Sanchez L (2015b) Genomic selection prediction accuracy in perennial crops: case study of oil palm (*Elaeisguineensis* Jacq.). Theor Appl Genet 128:397–410

Gaertner J (1788) De fructibus et seminibus plantarum, Stuttgart

Henderson J, Obsorne DJ (2000) The oil palm in our lives: how this came about. Endeavour 24:63–68

Henson IE (2012) A brief history of the oil palm. In: Lai O-M, Tan C-P, Casimir C (eds) Palm oil, production, processing, characterisation and uses. AOCS Press, Elseiver Inc, pp 1–29

Jacquin NJ (1763) Selectarum stirpium Americanarum historia. Wenen, Austria

Jones LH (1974) Propagation of clonal oil palm by tissue culture. Planter 50:374–381

Kushairi A, Loh SH, Azman I, Elina H, Ong-Abdullah M, Mohd Noor Izuddin ZB, Razmag G, Shamala S, Parveez A (2018) Oil palm economic performance in Malaysia and R&D progress in 2017. J Oil Palm Res 30(2):163–195

Kwong QB, Teh CK, Ong AL, Heng HY, Lee HL, Mohamed M, Low JZ-B, Sukganah A, Chew FT, Mayes S, Kulaveerasingam H, Tammi M, Appleton DR (2016) Development and validation of a high density SNP genotyping array for African oil palm. Mol Plant 9:1132–1141

Kwong QB, Ong AL, Teh CK, Chew FT, Tammi M, Mayes S, Kulaveerasingam H, Yeoh SH, Harikrishna JA, Appleton DR (2017) Genomic selection in commercial perennial crops: applicability and improvement in oil palm (*Elaeisguineensis* Jacq.). Sci Rep 7(1):2872

Maizura I, Rajanaidu N, Zakri AH, Cheah SC (2006) Assessment of genetic diversity in oil palm (Elaeisguineensis Jacq.) using Restriction Fragment Length Polymorphism (RFLP). Genet Resour Crop Evolut 53:187–195

Maria M, Clyde MM, Cheah SC (1995) Cytological analysis of *Elaeisguineensis* (tenera) chromosomes. Elaeis 7:122–143

Maria M, Clyde MM, Cheah SC (1998) Cytological analysis of *Elaeisguineensis*and*Elaeis oleifera* chromosomes. J Oil Palm Res 10(1):68–91

Obahiagbon FI (2012) A review: aspects of the African Oil Palm (*Elaeisguineesis* Jacq.). Am J Biochem Mol Biol 2(3):1–14

Ong-Abdullah M, Ordway JM, Jiang N, Ooi S-E, Kok S-Y, Sarpan N, Azimi N, Tarmizi Hashim A, Ishak Z, Samsul KR, Malike FA, Abu Bakar NR, Marjuni M, Abdullah N, Yaakub Z, Amiruddin MD, Nookiah R, Singh R, Low E-TL, Chan KL, Azizi N, Smith SW, Bacher B, Budiman MA, Van Brunt A, Wischmeyer C, Beil M, Hogan M, Lakey N, Lim C-C, Arulandoo X, Wong C-K, Choo C-N, Wong W-C, Kwan YY, Syed Alwee SSR, Sambanthamurthi R, Martienssen RA (2015) Loss of Karma transposon methylation underlies the mantled somaclonal variant of oil palm. Nature 525:533–537

Ong-Abdullah M, Ooi SE, Hashim AT, Ishak T, Rosli SK, Wong WC, Choo CN, Kok SY, Azimi N, Sarpan N (2017) 8.3: Revolutionising OPTC research. In: Soh AC, Mayes S, JAR Roberts (eds) Oil palm breeding: genetics and genomics. CRC Press, Taylor and Francis Group, New York, pp 197–216

Paterson RRM, Kumar L, Taylor S, Lima N (2015) Future climate effects on suitability for growth of oil palms in Malaysia and Indonesia. Sci Rep 5:14457

Price Z, Dumortier F, Macdonald D, Mayes S (2002) Characterisation of copia-like retrotransposons in oil palm (*Elaeisguineensis* Jacq.) Theor Appl Genet 104 (5):860–867

Price Z, Schulman AH, Mayes S (2003) Development of new marker methods-an example from oil palm. Plant Genet Resour 1(2–3):103–114

Rabechault H, Martin JP (1976) Multiplication vegetative de palmier a huile (*Elaeisguineensis* Jacq). C.R. Acad Sci Paris Ser D 283:1735–1737

Raymond WD (1961) The oil palm industry. Trop Sci 3:69

Rosenquist EA (1986) The genetic base of oil palm breeding populations. In: Soh AC, Rajanaidu N, Nasir M (eds) International workshop on oil palm germplasm and utilization. Palm Oil Research Institute of Malaysia, Kuala Lumpur, pp 16–27

Singh R, Ong-Abdullah M, Low E-TL, Manaf MAA, Rosli R, Nookiah R, Ooi LC-L, Ooi S-E, Chan K-L, Halim MA, Azizi N, Nagappan J, Bacher B, Lakey N, Smith SW, He D, Hogan M, Budiman MA, Lee EK

DeSalle R, Kudrna D, Goicoechea JL, Wing RA, Wilson RK, Fulton RS, Ordway JM, Martienssen RA, Sambanthamurthi R (2013a) Oil palm genome sequence reveals divergence of interfertile species in Old and New worlds. Nature 500:335–339

Singh R, Low E-TL, Ooi LC-L, Ong-Abdullah M, Ting N-C, Nagappan J, Nookiah R, Amiruddin MD, Rosli R, Manaf MAA, Chan K-L, Halim MA, Azizi N, Lakey N, Smith SW, Budiman MA, Hogan M, Bacher B, Van Brunt A, Wang C, Ordway JM,, Sambanthamurthi R, Martienssen R (2013b) The oil palm SHELL gene controls oil yield and encodes a homologue of SEEDSTICK. Nature 500:340–344

Singh R, Low E-TL, Ooi LC-L, Ong-Abdullah M, Nookiah R, Ting N-C, Marjuni M, Chan P-L, Ithnin M, Manaf MAA, Nagappan J, Chan K-L, Rosli R, Halim MA, Azizi N, Budiman MA, Lakey N, Bacher B, Van Brunt A, Wang C, Hogan M, He D, MacDonald JD, Smith SW, Ordway JM, Martienssen RA, Sambanthamurthi R (2014) The oil palm *VIRESCENS* gene controls fruit colour and encodes a R2R3-MYB. Nat Commun 5(4106):4106

Soh AC, Mayes S, Roberts JAR (2017a) Introduction to the oil palm crop. In: Soh AC, Mayes S, JAR Roberts (eds) Oil palm breeding: genetics and genomics. CRC Press, Taylor and Francis Group, New York, pp 1–6

Soh AC, Mayes S, Roberts JAR, Cros D, Purba R (2017b) Breeding plans and selection methods. In Soh AC, Mayes S, JAR Roberts (eds) Oil palm breeding: genetics and genomics. CRC Press, Taylor and Francis Group, New York, pp 229–236

Soh AC, Mayes S, Roberts JAR, Mahamouth T, Murphy DJ, Walker S, Karunaratne AS, Murchie E, Foulkes J, de Raissac M, Perez R, Fabre D, Goh KJ, Ong KO, Corley RHV (2017c) Future prospects. In: Soh AC, Mayes S, JAR Roberts (eds) Oil palm breeding: genetics and genomics. CRC Press, Taylor and Francis Group, New York, pp 229–236

Statistica, https://www.statista.com/statistics/263978/global-vegetable-oil-production-since-2000–2001/. Accessed 27 Aug 2018

Stilliard NH (1938) The rise and development of legitimate trade with West Africa. University of Birmingham, Thesis

Syed RA (1982) Insect pollination of oil palm: feasibility of introducing *Elaeidobius* spp. into Malaysia. In: Pushparajah E, Chew PS (eds) The oil palm in agriculture in the eighties, vol 1. Incorporated society of planter, Kuala Lumpur, pp 263–290

Teh HF, Mebus K, Neoh BK, Ooi TEK, Wong YC, Kulaveerasingam H, Appleton D (2017a) Omic technologies towards improvement of oil yield. In: Soh AC, Mayes S, JAR Roberts (eds) Oil palm breeding: genetics and genomics. CRC Press, Taylor and Francis Group, New York, pp 256–263

Teh CK, Kwong QB, Ong AL, Mohamed M, Apparow S, Chew FT, Mayes S, Appleton DR, Kulaveerasingam H (2017b). Application of genomic tools in oil palm breeding In: Soh AC, Mayes S, JAR Roberts (eds) Oil palm breeding: genetics and genomics. CRC Press, Taylor and Francis Group, New York, pp 246–256

Vijay V, Pimm SL, Jenkins CN, Smith SJ (2016) The impacts of oil palm on recent deforestation and biodiversity loss. PLoS ONE 11(7):e0159668. https://doi.org/10.1371/journal.pone.0159668

Wong CK, Bernado R (2008) Genome-wide selection in oil palm: increasing selection gain per unit time and cost with small populations. Theor Appl Genet 116:815–824

Zaki MN, Madon M, Schwarzacher Tand Heslop-Harrison JSP (2017) Chromosomes, cytology and molecular cytogenetics in oil palm. In: Soh AC, Mayes S, JAR Roberts (eds) Oil palm breeding: genetics and genomics. CRC Press, Taylor and Francis Group, New York, pp 229–236

Zeven (1964) On the origin of the oil palm. Grana Palynol 5:50

Wild and Advanced Resources of *Elaeis guineensis* and *Elaeis oleifera*

2

Mohd Din Amiruddin, Nor Azwani Abu Bakar, Fadila Ahmad Malike, Suzana Mustaffa, Norziha Abdullah, Marhalil Marjuni, Mohd Mustakim Mohamad, Mohd Yazid Hassan, Ahmad Kushairi, and Rajanaidu Nookiah

Abstract

The crucial role of genetic resources in agriculture cannot be taken lightly. These resources are required by plant breeders to improve and develop new crop varieties. Most plant commodities including modern varieties descended from an array of wild and improved genetic resources from around the world. Moreover, agricultural production depends on continuing infusions of genetic resources for yield stability and growth. In oil palm, the incipience of the wild resource in *Elaeis guineensis* was from West and Central Africa; meanwhile, the other species (*Elaeis oleifera*) was from America before these two species spread out globally and became important industrial crop. The breeding and genetic improvement work for oil palm began in the 1920s in Africa and later in Southeast Asia followed with the development of advanced cultivars of the crop from few ancestral palms. A significant improvement has been made from one generation to the next. However, as the crop improved, genetic diversity was reduced, and this restricted the ability to increase its productivity as well as selection progress further. The awareness of the depleting and narrowness of oil palm genetic diversity has become the impetus for the oil palm germplasm collection programme. Exploration of the oil palm genetic resources has been carried out by various countries to their centres of origin in Africa and Latin America. The primary purpose was to increase the genetic diversity for yield progress. The germplasm which possesses high levels of genetic diversity were collected, evaluated for valuable traits, utilized, conserved and acted as a reservoir of genetic diversity for future use. Utilization of both wild and advanced oil palm materials has expanded through various ways, which increases the potential benefits for crop improvement.

2.1 Introduction

There are two species of oil palm in the world, *Elaeis guineensis* and *E. oleifera*, the African and American oil palm, respectively. *E. guineensis* exists in a wild, semi-wild or cultivated state in three main areas of the equatorial tropics: Africa, Southeast Asia and South and Central America. Most of this spread has been a result of its domestication by man. Extensive natural or semi-

M. D. Amiruddin (✉) · N. A. Abu Bakar · F. Ahmad Malike · S. Mustaffa · N. Abdullah · M. Marjuni · M. M. Mohamad · M. Y. Hassan · A. Kushairi · R. Nookiah
Advanced Biotechnology and Breeding Centre, Malaysian Palm Oil Board, No. 6, Persiaran Institusi Bandar Baru Bangi, 43000 Kajang, Selangor, Malaysia
e-mail: mohddin@mpob.gov.my

© Springer Nature Switzerland AG 2020
M. Ithnin and A. Kushairi (eds.), *The Oil Palm Genome*, Compendium of Plant Genomes, https://doi.org/10.1007/978-3-030-22549-0_2

wild palm groves are distributed along the West Coast of Africa from Senegal to Angola (Zeven 1967). The second species, *E. oleifera*, originated in South and Central America and has been taken to other continents for breeding and research purposes. Wild *E. oleifera* can be found in countries such as Honduras, Nicaragua, Costa Rica, Panama, Colombia, Venezuela, Suriname, Ecuador and Brazil (Meunier 1975; Ooi et al. 1981; Escobar 1982; Rajanaidu 1983; Mohd Din 2000; Mohd Din et al. 2000). From genome sequences, Singh et al. (2013) estimated that the two species of *Elaeis* diverged for as long as 51 Myr ago. Despite this, the two species can form more or less fertile hybrids.

The oil palm commercial planting materials are known as *tenera*. *Tenera* (T) is a hybrid seed produced from crossing between thick-shelled *dura* (D) and a shell-less *pisifera* (P). The *dura* and *pisifera* individuals are evaluated and selected separately according to a procedure called modified recurrent selection (MRS). In MRS method, potential *dura* and *pisifera* parental palms are intercrossed and carried forward to the next cycle for future breeding and selection. These palms are also hybridized to evaluate progeny performance and combining ability. *Dura* and *pisifera* palms that produced outstanding progenies and recorded good combining ability are selected and used for commercial seed production.

2.2 Wild Resources of Oil Palm

History has shown the importance of having a wide genetic base for crop improvement. It has been generally recognized that the narrowness of gene pools has been a major obstacle to selection progress in oil palm. Further improvements were foresighted and executed in the widening of the genetic pool and collections of germplasm in centres of origin/diversity in Africa and Latin America. Major oil palm germplasm explorations were carried out by Malaysian Palm Oil Board (MPOB), Indonesian Palm Oil Board (IPOB) and Colombian Oil Palm research Centre (CENIPALMA).

2.2.1 Malaysian Oil Palm Germplasm Collection

Malaysia has collected *Elaeis guineensis* germplasm from countries in Africa such as Nigeria, Cameroon, Zaire, Tanzania, Madagascar, Senegal, Gambia, Sierra Leone, Guinea, Ghana and Angola. *Elaeis oleifera* germplasm, on the other hand, was collected in Honduras, Nicaragua, Costa Rica, Panama, Colombia, Suriname and Ecuador (Meunier 1975; Ooi et al.1981; Escobar 1982; Rajanaidu 1983; Rajanaidu and Rao 1987; Mohd Din 2000; Mohd Din et al. 2000; Rajanaidu et al. 2013a, b; Rajanaidu et al. 2017) (Table 2.1).

The germplasm collections are maintained in field collection for posterity. In order to ensure effective use, the oil palm germplasm has been evaluated for oil palm improvement programmes. As reported by Rajanaidu et al. (2017), among the *Elaeis guineensis* germplasm collections, Nigeria had the highest mean bunch weight for *duras* and *teneras* (24–26 kg) compared to the other germplasm. Both Nigeria and Angola had 80% mesocarp to fruit, which is the highest among the germplasm. Tanzanian materials had fruit qualities comparable to those collected in Nigeria, Cameroon, Zaire and Angola. In addition, Tanzanian also had the highest fruit weight for both *duras* and *teneras* with 9 and 15 g, respectively.

General comparison made among the germplasm materials collected from Central and South America for *Elaeis oleifera* showed that the highest mean bunch weight was recorded by palms from Colombia germplasm (41 kg). As for the mesocarp contents, individual palms from Colombia exhibited the highest (65.46%). In terms of fruit size, bunches sampled along the Atlantic Coast had bigger fruits than those in Pacific Coast. In addition, the mesocarp content in Central America was lower than those in South America, Brazil and Kuala Lumpur Melanococca (KLM) palms (Rajanaidu et al. (2017).

Selected palms that revealed outstanding phenotypic traits were distributed to oil palm private companies in Malaysia, namely United

Table 2.1 List of MPOB germplasm collection mounted in both Africa (*E. guineensis*), Central and South America (*E. oleifera)* from 1973 to 2010

Country	Year	Collectors	No of accessions (based on fruit type)			Total no of accessions collected
			D	T	P	
Elaeis guineensis						
Nigeria	1973	Rajanaidu, Arasu and Obasola	595	324	0	919
Cameroon	1984	Rajanaidu and Unilever	53	42	0	95
Zaire	1984	Rajanaidu and Unilever	284	85	0	369
Tanzania	1986	Rajanaidu and Ministry of Agriculture Officials, Tanzania	42	17	0	59
Madagascar	1986	Rajanaidu and Ministry of Agriculture Officials, Madagascar	17	0	0	17
Angola	1991	Rajanaidu, Jalani Sukaimi, Mohd Din Amiruddin and Ministry of Agriculture Officials, Angola	42	12	0	54
Angola	2010	Rajanaidu, Marhalil Marjuni and Ministry of Agriculture Officials, Angola	95	32	0	127
Senegal	1993	Rajanaidu, Jalani Sukaimi and Ministry of Agriculture Officials, Senegal	104	0	0	104
Gambia	1993	Rajanaidu, Jalaini Sukaimi and Ministry of Agriculture Officials, Gambia	45	0	0	45
Sierra Leone	1994	Rajanaidu, Ahmad Kushairi and Ministry of Agriculture Officials, Sierra Leone	52	3	1	56
Guinea	1994	Rajanaidu, Ahmad Kushairi and Ministry of Agriculture Offcials, Guinea	58	3	0	61
Ghana	1996	Rajanaidu, Mohd Rafii Yusop and Ministry of Agriculture Officials, Ghana	48	10	0	58
Elaeis oleifera						
Honduras	1982	Rajanaidu	14			14
Nicaragua	1982	Rajanaidu	18			18
Costa Rica	1982	Rajanaidu	61			61
Panama	1982	Rajanaidu	27			27
Colombia	1982	Rajanaidu	42			42
Suriname	1982	Rajanaidu	7			7
Ecuador	2004, 2006	Rajanaidu, Ahmad Kushairi and Noh Ahmad	11			11

Total number of accessions for both *E. guineensis* and *E. oleifera* in MPOB genebank are close to 2000

Plantations Berhad (UPB), Sime Darby Plantation, Federal Land Development Authority (FELDA), Kulim Plantations Berhad, IJM Plantation Berhad and Applied Agricultural Resources (AAR). Close collaborations were established between MPOB and these companies to evaluate crosses derived from the germplasm materials and advanced breeding populations.

2.2.2 NIFOR Oil Palm Germplasm Collection

In Nigeria, the main prospection was undertaken by NIFOR (Ataga et al. 1999). The first exploration was mounted in Southeastern Nigeria involving Calabar, Aba, Nkwele in 1912 and, later, Ufuma (Okwuagwu 1986). Besides Nigeria, due to the liberal genetic exchange policy, NIFOR also acquired genetic materials from other African countries, Southeast Asia and Latin America (Table 2.2).

2.2.3 Indian Germplasm Collection

Department of Agriculture, Kerala was the leading organization in collecting the oil palm germplasm materials in 1960s in India. The Department imported genetic materials from Malaysia and Nigeria in the form of hybrid seeds (D×D, D×T, D×P and T×T) and planted at Oil Palm Station at Thodupuzha, Kerala (Nampoothiri and Pillai 1998). In 1979, active collection of oil palm accessions was carried out by the Indian Council of Agricultural Research. The ex situ field gene banks consisting of collections from 11 countries have been established at National Research Centre for Oil Palm, Pedavegi (Andhra Pradesh), and Research Centre of NRC for Oil Palm, Palode (Kerala) (Auxilia and Shabha 2017). Some cold-tolerant oil palm materials from African countries, namely Guinea Bissau, Cameroon, Zambia and Tanzania, were collected by Central Plantation Crop Research Institute (CPCRI) in collaboration with Food and Agriculture Organization (FAO) in 1993. These collections were tested for yield performance and stress tolerance in seven areas in India, namely Palode, Power Company of Karnataka Ltd (PCKL), Pedavegi, Adilabad, Nellore, Mulde and Mohitnagar (Pillai et al. 2000). Table 2.3 presents some oil palm genetic materials assembled from various countries that

Table 2.2 NIFOR oil palm germplasm collection

Origin	Country	Organization	Year of collection
Calabar	Nigeria	NIFOR	1912–1916
Aba	Nigeria	NIFOR	1939–1941
Ufuma	Nigeria	NIFOR	1939
Angola	Angola	Siera Leone	1960s
Yangambi	Zaire	INEAC	1960s
Binga	Congo	Research Department, Binga	1960s
Lobe	Cameroon	Research Department	1960s
La Mé	Ivory coast	IRHO	1960s
Sabah Deli	Malaysia	DOA, Malaysia	1960s
Serdang Avenue Deli	Malaysia	DOA, Malaysia	1948
Ulu Remis Deli	Malaysia	Chemara Programme	1960s
Dabou Deli	Ivory Coast	IRHO	1950s
Pobe Deli	Ivory Coast	IRHO	1950s
Ecuador	Ecuador	INIAP	1960s
Jamaica	Jamaica	–	1950s

Table 2.3 List of countries covered by the Indians to assemble oil palm germplasm collection

	Country	Accessions
1	Sierra Leone	EC36788
2	Nigeria	EC39991–92
3	Malaysia	EC66175
4	Czech Republic	EC95232
5	Malaysia	EC99352
6	Nigeria	EC130657
7	Canada	EC159876–77
8	Cameroon	EC188427
9	Singapore	EC191687–94
10	Philippines	EC232497
11	Costa Rica	EC332236–304
12	Cameroon	EC375360–76
13	Belgium	EC376297–321
14	Tanzania	EC382627–37
15	Zambia	EC382638–46
16	Costa Rica	EC453781–82
17	Costa Rica	EC454569–70

are maintained by the National Bureau of Plant Genetic Resources (NBPGR), New Delhi (Pedapati et al. 2013).

2.2.4 Indonesian Oil Palm Germplasm Collection

Indonesia's first exploration for oil palm germplasm officially commenced in 2008 in Cameroon. A total of 103 accessions were collected (Bakoumé 2016). The expedition was mounted in collaboration with researchers from the Indonesian Palm Oil Board (IPOB) and the Cameroon Institute of Agricultural Research for Development (IRAD). In 2010, 127 oil palm accessions were collected in Angola through cooperation with the Malaysian Palm Oil Board (MPOB) (Rajanaidu and Ainul 2013). These accessions revealed various fruit colours such as *nigrescens*, *virescens* and *albescens*. The genetic materials also exhibited valuable traits such as compactness, long stalk, big fruits, short fronds and thick mesocarp. At the end of 2016, Indonesia took the initiative to purchase commercial *oleifera* × *guineensis* hybrid seeds from Colombia and Ecuador.

2.3 Outstanding Traits Found in the Germplasm Collection

MPOB oil palm germplasm were planted in the form of 'open-pollinated' families at MPOB Kluang Research Station, Johor. Many years have been spent on the evaluation of performance of individual palm based on cubic lattice, completely randomized (CRD) and randomized complete block (RCBD) experimental designs. Detailed list of traits studied is reported by Rajanaidu et al. (2000). It takes approximately 10 years of field evaluation before palms with interesting traits can be selected and offered to the Malaysian oil palm industry as Planting Series (PS) for incorporation into their breeding and improvement programmes.

Short palm programme has become a priority in the Malaysian oil palm industry since the

1960s due to difficulty in harvesting bunches from matured oil palms. At growth rate of 40–75 cm/year, the current planting materials easily reach a height of 10 metres after 15 years. Initially, the short palm programme made use of the slow height increment *oleiferas* through inter-specific hybrid breeding programmes (Hardon and Tan 1969). Selected *oleifera* palms that recorded growth rate of 4.6 cm/year (Mohd Din et al. 2000) were hybridized with *E. guineensis*. However, the hybrids turned out to be low yielding and less promising in terms of oil to bunch.

The first generation of value-added planting materials developed from germplasm populations is PS1 or Planting Series 1. PS1 was created by crossing high fresh fruit bunch (FFB) and dwarf Nigerian *dura* palms (population 12) with selected advanced AVROS *pisiferas*. The selected Nigerian *dura* palms possessed high FFB yield more than 200 kg/p/year and height increment between 15 and 19 cm/year. In addition, selected Nigerian dwarf *teneras* were also selfed and intercrossed to produce *pisiferas* that were distributed to the Malaysian oil palm companies for progeny testing with their Deli *duras*. These derived dwarf *pisiferas* were selected to enable production of larger number of crosses (Kushairi et al. 2001).

The Nigerian germplasm also exhibited iodine value (IV) of more than 60 compared to the commercial materials (IV = 50–53). These palms were introduced to the oil palm industry as PS2. Some of the PS2 palms were progeny tested with AAR Yangambi AVROS and Dumpy-AVROS *pisiferas* resulting in higher IV (>58.0) but lower yield (126 and 165 kg/p/yr) compared to D×P standard cross (160 and 216 kg/p/yr) (Sharma 1999).

The Nigerian germplasm collection offers valuable traits for development of oil palm with high lauric oil, an important raw material in oleochemical industry. Selected Nigerian palms, known as PS3, exhibited 10–15% of kernel content (Rajanaidu et al. 2000) compared to 5–7% of the commercial planting materials. PS12 is another Planting Series that possessed oleic acid content between 48 and 52% (Isa et al. 2006) which is higher than the current planting materials (37–40%). These materials are potentially useful for development of high lauric oil palm planting materials.

Oil palm breeding programmes are mainly centred on improvement of oil yield. Evaluation of the Tanzania collection revealed five *tenera* palms with low shell-to-fruit ratio (2.80–7.40%) as compared to 12% in the commercial materials. These palms were released to the industry as PS5 (Kushairi et al. 2003a). PS6, which possessed large fruit *dura* size (24–34 g) was discovered from the Angola germplasm populations (Kushairi et al. 2003b). High bunch index palms or PS7 were identified from Tanzania, Cameroon, Nigeria and Angola germplasm collections. Palms with high bunch index also have high FFB yield. Incorporating PS5, PS6 and PS7 into breeding programmes could potentially improve palm oil yield.

The oil palm germplasm collections also offer valuable traits for development of new planting materials with diversified products. Palms with high carotene content were identified in both *E. oleifera* (named PS4) and *E. guineensis* (known as PS11) germplasm collections. High carotene *E. oleifera* possessed more than 3000 ppm, whereas in *E. guineensis*, the carotene content ranged from 2000 to 2474 ppm. These are higher than that recorded for the commercial planting materials (500–700 ppm). Selected germplasm palms recorded vitamin E between 1300 and 2500 ppm, which is higher than the commercial materials (600–1000 ppm). These palms were introduced as PS8 (Kushairi et al. 2004) for improvement of vitamin E content in the current planting materials for pharmaceutical application. The current planting materials exhibited 16% protein in the kernel. Some germplasm palms from Nigeria, Angola, Zaire, Tanzania, Cameroon, Guinea, Sierra Leone, Senegal, Gambia and Ghana showed more than 20% protein in the kernel (Noh et al. 2005). These palms can be incorporated into breeding programmes to develop new oil palm that produces higher protein content in the kernel. This high protein kernel is cheaper than other raw feed ingredients for feedstuff production. The lower cost of feedstuff also means lower cost of livestock production.

In terms of productivity, planting materials that produce bunches with longer stalk (PS10) can increase harvesting efficiency. Ten Angola palms that exhibited longer stalk (20–30 cm) (Noh et al. 2005) as compared to the present commercial materials (10–15 cm) were identified. Bunches with longer stalk are easier to harvest, thus increasing workers' productivity. Low lipase palms or PS13 revealed lipase activity between 1 and 10% compared to commercial planting materials (22–73%) (Maizura et al. 2008). Palms with low lipase bunches produce better quality palm oil that fetches higher price. Some Tanzania *dura* materials showed low height increment (<0.30 m/year) and shorter rachis (<5.0 m) (Marhalil 2009). These materials are being tested for the development of compact palm.

2.4 Performance of Progeny Test Experiments Involving Germplasm Materials

2.4.1 Performance of Introgressed Progeny Derived from MPOB Germplasm

The genetic variability of the current breeding populations is broadened through introgression of advance materials with selected materials from germplasm maintained by MPOB. The Nigerian population 12 materials (PS1) have been introgressed into oil palm breeding programmes in Malaysia since the 1980s (Marhalil et al. 2017). Selected PS1 *pisiferas* palms were crossed with high-quality Deli *duras*. The main focus is to transfer the dwarf and high yielding traits of population 12 without compromising fruit qualities of the Deli *dura*. The mean performance of the PS1 progenies showed that the grand mean for FFB yield was 27.31 t/ha/yr and the best progeny yielded 28.76 t/ha/yr with oil yield at 8.19 t/ha/yr (Table 2.4). The best oil-to-bunch ratio (O/B) attained was 29.41% with oil extraction rate at 25%. The lowest height increment achieved was 31.33 cm/yr (Marhalil et al. 2016). These results indicated the potential use of the selected Nigerian palms for improvement of yield and height of the commercial planting materials.

Further evaluation of Nigerian introgressed populations was carried out by Arolu et al. (2016). Some progenies (DP5, DP8, DP4, DP3 and DP24) possessed high yielding and dwarf palms (Table 2.5). At planting density of 132 palm/ha, the progenies showed FFB between 29.33 and 32.51 t/ha/yr, average 27.33 t/ha/yr which is about 16.7% higher than the current planting material (t/ha/yr). Selected progenies possessed average annual height increment of 36.67 cm/yr, ranged between 29.17 and 46.33 cm/yr. These are 22% shorter than that recorded for the current planting material (45–75 cm/yr).

Table 2.4 Performance of D×P progeny test (Male) of new introduction Nigerian *pisifera* in Kota Tinggi, Johor

No	Male	Crosstype	FFB (t/ha/yr)	ABW (kg)	MFW (g)	M/F (%)	S/F (%)	O/DM (%)	O/B (%)	OY (t/ha/yr)	HTi (cm/yr)
1	P01	D×P	28.76	11.2	10.77	83.21	11.57	79.32	28.28	8.19	39.83
2	P02	D×P	26.41	11.92	12.96	80.96	12.52	79.01	26.61	6.93	35.50
3	P03	D×P	27.79	11.99	11.57	84.61	10.69	81.11	29.41	8.04	38.50
4	P04	D×P	26.60	12.79	15.28	80.91	12.39	81.09	27.63	7.38	31.33
5	P05	D×P	26.97	11.32	11.3	83.62	10.592	80.86	28.58	7.73	32.83
Mean			27.31	11.84	12.38	82.67	11.55	80.28	28.10	7.65	35.67

Notes Planted in 2004
FFB fresh fruit bunches, *ABW* average bunch weight, *MFW* mean fruit weight, *M/F* mesocarp to fruit ratio, *S/F* shell to fruit ratio, *O/DM* oil to dry mesocarp, *O/B* oil to bunch ratio, *OY* oil yield, *HTi* height increment

Table 2.5 Progeny performance of Nigerian-based DxP (new introduction Nigerian *pisifera*) at Kota Tinggi, Johor

Code	Progeny	FFB (t/ha/yr)	BNO	ABW (kg)	HTi (cm/yr)	MFW (g)	MTF (%)	OTB (%)
DP1	ECPHP105	31.25	19.12	11.09	37.50	9.90	85.22	31.00
DP2	ECPHP108	31.23	18.96	11.30	34.67	11.48	82.41	27.54
DP3	ECPHP110	32.51	20.25	11.01	43.00	9.78	84.02	29.42
DP4	ECPHP130	31.44	17.60	12.24	43.33	12.10	86.09	30.28
DP5	ECPHP200	29.47	17.55	11.50	33.17	14.24	79.79	25.19
DP6	ECPHP208	30.08	16.67	12.38	29.17	14.05	81.98	27.88
DP7	ECPHP218	29.50	16.18	12.55	37.33	13.32	80.00	26.81
DP8	ECPHP230	30.27	15.54	13.41	40.33	14.07	81.36	26.96
DP9	ECPHP256	29.78	14.98	13.53	35.17	15.79	80.27	28.34
DP10	ECPHP347	29.33	15.92	12.54	33.33	10.10	83.87	28.21
DP11	SC PK3833	25.41	13.96	12.45	46.33	14.57	78.85	28.32
Trial mean		30.51	17.28	12.16	36.67	12.48	82.50	28.16

Notes FFB fresh fruit bunch, *BNO* bunch number, *ABW* average bunch weight, *HTi* height increment, *MFW* mean fruit weight, *OTB* oil to bunch, *SC* standard cross

Oil palm genetic improvement and sustainable development take numerous criteria into account such as production of high yield, resistance to the main diseases (vascular wilt, Ganoderma, bud rot), drought tolerance and material with low fertilizer requirement. Genetic diversity must be maintained not only when creating new variety but also when establishing commercial areas (Cochard et al. 2005). Efforts to diversify the current *dura* genetic pool were initiated by introducing Nigerian *dura* palms into crossing schemes. The resulting *duras* were hybridized with the established *pisiferas* that were extensively used for producing commercial oil palm planting materials (Alwee et al. 2017b). Alwee et al. (2017a) reported on exploiting germplasm materials for drought tolerance. A fast screening system was developed at nursery stage to evaluate the response of various progenies towards drought conditions. Progenies that showed tolerance to dry conditions in the nursery will be further evaluated in Chuping, Perlis, Malaysia, and is classified as one of the driest areas in the country.

2.4.2 Germplasm Introgression Programme of Other Organizations

In Nigeria, the genetic improvement efforts of oil palms started with the establishment of the Nigerian Institute for Oil Palm Research (NIFOR). At present, the oil palm breeding programme undertaken by NIFOR is focusing on testing oil palm genotypes for high yield and adaptation to different agro-ecological areas. Okoye et al. (2018) reported differences in genotype clustering based on agronomic and molecular analysis revealed, even though they shared several common aspects, such as high diversity between DT9 (Ufuma × Umaibi OP, D×P) and DT6 (Ecuador *deli* × Calabar, D×P) genotypes. These genotypes had been previously selected and evaluated in NIFOR breeding programme and exhibited good performance. DT9 population recorded the highest average bunch weight production with the least coefficient of variance (CV), denoting the uniformity of the genotype for this agronomic trait.

In India, majority of the oil palm germplasm materials conserved were collected from different countries aiming to ameliorate narrow genetic base and produce superior dwarf hybrids suitable to different agro-climatic conditions and tolerant to abiotic stress (Pedapati et al. 2013). A comparative trial of different oil palm hybrids was conducted at the oil palm centres of All India Coordinated Research Project on Palms in different agro-climatic regions. Results showed that hybrid 124D×266P recorded high FFB yield of 24.06 tonnes/ha at Mulde centre in Maharashtra. In addition, hybrid 115D×291P which was planted at Vijayarai and Mulde revealed FFB of 13.80 tonnes/ha and 20.32 tonnes/ha, respectively. Evaluation of drought-tolerant *dura* from the oil palm germplasm was carried out under rain-fed conditions at Gangavathi (Karnataka) and Mulde (Maharashtra) centres. At Gangavathi centre, significant differences were observed in lipid peroxidation levels indicating drought tolerance (Arulraj and Suresh 2011).

The breeding objectives of the Indonesian Oil Palm Research Institute (IOPRI) are to increase oil yield, develop disease resistance variety as well as improve oil quality. A huge scale of screening against Ganoderma was carried out, involving 350 D×P hybrid crosses. Efforts of improving oil quality were also initiated through backcross breeding programme involving *oleifera* and *guineensis* palms. In addition, introduction of new oil palm genetic materials from Cameroon and Angola has opened up the opportunity for IOPRI to develop oil palm cultivars with novel traits (Suprianto et al. 2016)

2.5 Advanced Breeding Resources of Oil Palm

2.5.1 The Deli *Dura* Breeding Population

The oil palm industry in the Far East owes its beginning to four *dura* seedlings introduced at the Bogor Botanical Garden, Indonesia, in 1848 (Hartley 1988). Seeds from Bogor were planted for esthetical purposes along avenues in Sumatra, notably in Deli, in the 1870s. Materials from Sumatra were imported to Malaysia on several occasions, but those planted along avenues in Rantau Panjang, Selangor, in the 1910s mooted the oil palm industry in Malaysia. It should be realized that while acquiring open-pollinated seeds for avenue plantings, and the subsequent selection to supply seeds for more avenues, superior materials were unconsciously mass selected for several generations. Formal breeding and selections were carried out since the 1920s in Indonesia at Marihat Baris Estate, and in Malaysia at Elmina Estate and Serdang Experimental Station (FES) (Hardon and Thomas 1968). After intense breeding and selections, these palms had become uniform in performance, yielding big bunches, good fruit characteristics with high mesocarp content. This population became known as the Deli *dura* population. Subsequent imported *duras* from Africa did not match the fruit qualities of the Deli *dura*. The Dumpy *dura* E206 is a mutant of Deli *dura* identified in Elmina Estate by Jagoe (1952). Despite the low bunch yield that did not match the selection criteria, palm E206 was selected for its short stature. It is characterized by short palm height, large girth, large petiole, big bunches and good fruit characteristics. The Dumpy *dura* and its derivatives are featured in several breeding programmes worldwide in a renewed interest in planting shorter palms for the ease of harvesting.

The oil palm breeding populations have been developed from few ancestral palms. The basic population in major breeding programmes worldwide, with the probable exception of Nigeria, is almost exclusively derived from the Deli *dura*. Although a breeding population may have been introgressed with other populations, there still remain a considerable number of populations derived from a few ancestral palms. One programme has sometimes overlapped with another, but Rosenquist (1986) considered the following as independent 'breeding populations of restricted origins' (BPRO):

Malaysia

- Serdang Avenue
- Elmina
- Ulu Remis Banting
- Johor Labis.

Indonesia

- Gunung Bayu
- Pabatu Blocks 87, 88
- Dolok Sinumbah–Tinjowan–RISPA
- Pabatu Block 54
- Marihat Baris
- Mapoli/Bangun Bandar
- Gunung Melayu.

Ivory Coast

- Dabou (Ex-Deli *dura*).

Zaire

- Lofindi via Yangambi and Lokutu.

Papua New Guinea

- Dami (Ex-Deli *dura* via MARDI, Guthrie Chemara, Banting.

Costa Rica

- Coto (Ex-Deli *dura* via MARDI, Guthrie Chemara, Banting).

The Department of Agriculture (DOA) pioneered the oil palm breeding programme in Malaysia in the 1920s, followed by two other companies, Guthrie Berhad and Societe Financiere de Caoutchouces (Socfin) in the 1930s (Kushairi and Rajanaidu 2000). Elmina (e.g. E152 and E206), Johor Labis (e.g. JL 4107 and JL 1407), Ulu Remis (e.g. UR 27.9 and UR 413/13) and some Socfin materials were some of the earliest *dura* parental materials used and the first generation of BPRO *duras*. These materials were planted between 1930 and 1952 in Federal Experimental Station (F.E.S), Fields 3A and 4 at Serdang, Selangor, and Jerangau, Terengganu.

The Dumpy E206 from Elmina was known for its low stature compared to 'normal' Elmina palms. The Dumpy families were also more uniform in yield and other characters as compared to the tall progeny lines. Meanwhile, Serdang Avenue *dura* palms showed higher yielding and kernel to fruit percentages than those of the commercial Deli (Hartley 1988). It also had higher oil-to-bunch (O/B) ratio than the Elmina *duras*. However, when crossed to AVROS pisifera, Serdang Avenue and Elmina produced progenies with comparable O/B (Rosenquist 1986). Through the performance records of Sabah Breeding Programme, Banting *duras* (introgession between Chemara and Serdang palms) showed high yields.

Dura improvement programme was carried out using the following sets of *dura*: Ulu Remis (UR (D)), Johor Labis (JL), Klanang Bahru (KB), Highlands Estate (HE, HEZE, HEZG), Banting (BD), Serdang Avenue (Serd.), Tumbuk Estate, Gunung Melayu (GM). These materials were subjected to breeding and selection processes and now have reached the fourth and fifth generation. In between, the materials were distributed to oil palm agencies under joint trial basis. The performance of the fourth and fifth generation is presented in Table 2.6. Selected *dura* from the fifth cycle has been used for commercial seed production (Marhalil et al. 2016).

Hardon et al. (1987) estimated that selection progress in the second and subsequent generations was about 10–15% per generation. Selection progress for fresh fruit bunch (FFB) yield of *duras* from the fourth to fifth cycle ranged from 12 to 16%, which corresponded with the prediction of Hardon et al. (1987). Other traits showed negative or least progress, suspected to be due to inbreeding depression (Marhalil et al. 2016)

Table 2.6 Mean performance of the fourth and fifth cycles of *dura* progenies

Origin	Cycle	FFB (t/ha/yr)	ABW (kg)	MFW (g)	M/F (%)	S/F (%)	O/DM (%)	O/B (%)
Banting	4th	11.12	10.51	9.13	59.91	33.03	76.46	18.93
	5th	12.49	11.47	10.54	64.25	28.74	71.69	18.18
Elmina	4th	14.00	13.38	12.23	60.16	32.05	75.90	17.17
	5th	16.19	14.45	12.84	62.10	30.94	74.32	18.12
Ulu Remis	4th	15.35	12.39	13.89	57.01	35.32	72.60	15.41
	5th	13.52	12.56	14.45	58.71	33.28	70.17	16.01

Notes Trial 0.212 and 0.400 were planted in 1986 and 2000, respectively (148 palms/ha)
FFB fresh fruit bunch, *ABW* average bunch weight, *MFW* mean fruit weight, *M/F* mesocarp to fruit, *S/F* shell to fruit, *O/DM* oil to dry mesocarp, *O/B* oil to bunch

2.5.2 The *Tenera/Pisifera* Breeding Populations

While the *dura* is exclusively based on the Deli population, the *tenera/pisifera* populations used in breeding programmes came from a wider range of sources. The *pisiferas* were generated from either *tenera* × *tenera* (T×T), *tenera* × *pisifera* (T×P) or *pisifera* × *pisifera* (P×P) crosses. Some of the common *pisiferas* in breeding and seed production are as follows:

- Yangambi
- AVROS
- Serdang
- NIFOR
- La Mé
- Derived *pisiferas*

Yangambi *pisifera*

The Yangambi *pisifera* originates from the breeding programme of the Institute National pour l'Etude Agronomique du Congo Belge (INEAC) in the Republic of Congo (Zaire). Before the Second World War, INEAC made several collections in the wild groves of Congo. Open-pollinated material from Eala, Yawenda, N'gazi and Isangi formed the Yangambi population. This population, notably the Djongo (best) palm, is extensively featured in major seed production and breeding programmes worldwide as described by Beirneart (1993) and Vanderweyen (1952). D×P progeny of Yangambi *pisifera* is characterized by vigorous growth, high early yields, ovoid fruit, thin shell and displaced kernel. The Yangambi population is widely distributed to many centres of the world and had evolved under various names, such as the Binga, AVROS and BM *pisiferas*.

AVROS *pisifera*

The AVROS *pisifera* is closely related to the Yangambi *pisifera* being a descendant of the Djongo palm in Eala as described by Rosenquist (1986). Seeds of the Djongo palm were imported to Sumatra by *Algemene Vereniging van Rubber planters ter Oostkust van Sumatra* (AVROS) and planted in Sungai Pancur in 1923. One of the palms planted was SP540 *tenera*. The SP540 palm was selfed and crossed with *teneras* from Bangun Bandar Experimental Station in Sumatra. After intense breeding and selection, the AVROS *pisifera* showed high values for general combining ability (GCA) and uniformity in their performances (Kushairi and Rajanaidu 2000). The resulting D×P progenies exhibit high mesocarp to fruit and oil-to-bunch ratios. Like the Yangambi *pisifera*, the progeny is also vigorous growing, precocious (early bearing), thin shell, good oil extraction and ovoid fruits, but with big kernels centrally placed. The AVROS *pisifera* is widely distributed throughout the world.

La Mé *pisifera*

The La Mé *pisiferas*, notably sibs of L2T, L5T and L7T, originated from the palms near the La

Mé Research Station in Ivory Coast. Breeding materials of the Institute de Resherches pourLes Huiles et Oleagineux (IRHO) were derived from 38 palms selected from Pobe in Benin and four palms selected at Bingerville in Ivory Coast. The latter provided the well-known La Mé population, the 'L2T' *tenera*. The selfed progenies of L2T have produced some outstanding *pisiferas*. It is also the parent of the standard cross L2T×D10D as mentioned by De Berchoux and Gascon (1965). The La Mé D×P progenies are characterized by short palm height, large number of small bunches with small fruits and kernel. Another distinguishing feature of La Mé *pisifera* is the longer bunch stalk. In practice, pollen of L2T *pisifera* is difficult to obtain because of its highly inbred nature.

Serdang *pisifera*

The Department of Agriculture (DOA), Malaysia, selected the Serdang *pisifera* (SP). The common ones being S29.36 and S27B. These *pisiferas* have large kernels, 0.7 and 0.4 cm diameter, respectively. Serdang *pisiferas* played significant roles in the breeding and selection programmes of many seed producers in Malaysia. Before the importation of tested *pisiferas*, Serdang *pisifera* was the major source of pollen for seed production in Malaysia. The D×P progenies of S27B are high in bunch number, while S29.36 confers better fruit characteristics. The D×P progenies arising from the Serdang *pisiferas* are shorter compared with those of Yangambi/AVROS, but taller than the La Mé-based population.

NIFOR *pisifera*

Pisifera populations of the Nigerian Institute for Oil Palm Research (NIFOR) have a broader genetic base than other programmes elsewhere. NIFOR *pisiferas* had been selected from germplasm collections such as from Calabar, Aba, Ufuma, Umuabi, Cowan Estate and Igala. Of these, D×P progenies derived from Calabar were the most interesting. Some examples of Calabar *pisiferas* are NF 5.1654 (or WA3), NF 6.594 (or WA9), NF 32.364, NF. 32.2612 and NF32.3005 (Sparnaaij et al. 1963). The NIFOR *pisiferas* had been distributed elsewhere, including Lobe in Cameroon and Sabah Breeding Programme (SBP) in Malaysia. Characteristics of the D×P progenies are not definitive because of the varied nature of the materials. Some progenies from Deli×NF showed good performance in progeny trials (Chin and Suhaimi 1999).

Derived *pisifera*

Besides individual palm selection, *pisifera* populations had also been developed through introgression of palms from various breeding populations. Some examples of derived *pisifera* are the Dumpy-AVROS and Ulu Remis *Tenera* (URT). While the URT was developed for yield improvements, the Dumpy-AVROS was generated with the aim of introgressing the low stature of the Dumpy *dura* with outstanding yield of the AVROS *pisifera*. It was noted (Kushairi et al. 1998) that progenies of *dura* × E206-based-*pisiferas* (i.e. E206 as the male parent) gave lower height increments compared with E206 *dura* × AVROS (i.e. E206 as the female parent) that gave normal palm heights.

AVROS from Trial 0.79 planted in F. E. S. Serdang was the most known and utilized *pisifera* source in MPOB. Potential progenies from this trial were then brought forward to succeeding generation and planted in Trial 0.174 and 0.182 at MPOB Kluang Research Station, Johor, Malaysia. The subsequent cycle was created from potential and promising AVROS through *tenera* × *pisifera* (T×P) crosses which were then planted in Trials 0.394 and 0.395 at MPOB Hulu Paka and MPOB Kluang Research Stations, respectively. The mean performance of AVROS *pisifera* is shown in Table 2.7 (Marhalil et al. 2016). Both AVROS *pisiferas* in trials 0.174 and 0.395 showed comparable results for most of bunch quality traits.

Table 2.7 Mean performance of *pisiferas* in trials 0.174 and 0.395

Trial	Cross	MFW (g)	M/F (%)	S/F (%)	O/DM (%)	O/B (%)	OY (t/ha/yr)
0.174	T×T	10.83	83.70	9.00	78.70	26.90	3.98
0.395	T×P	12.27	84.17	8.87	76.76	26.90	3.89

Notes Trial 0.174 and 0.395 were planted in 1981 and 2000, respectively (148 palms/ha) *MFW* mean fruit weight, *M/F* mesocarp to fruit, *S/F* shell to fruit, *O/DM* oil to dry mesocarp, *O/B* oil to bunch, OY = oil yield

2.5.3 Semi- and Bi-Clonal Planting Materials

Production of semi and bi-clonal seeds is an alternative method to produce uniform high yielding planting materials (Veerappan et al. 2000; Soh et al. 2005). The *dura* and *pisifera* palms that recorded good specific combining ability (SCA) are selected and cloned. The clonal parents are then crossed in the same way as that in conventional D×P seed production.

Semi-clonal and bi-clonal seeds have several advantages over the conventional D×P seeds in terms of degree of uniformity and oil yield gain. As the crossings involve limited number of parental combinations, the semi- and bi-clonal seeds are more uniform with an expectation of 15% of oil yield gain compared to conventional D×P hybrid seeds (Kushairi et al. 2010). United Plantations Berhad (UPB) in Malaysia pioneered the bi-clonal seeds production in the world with estimated annual production of one million D×P bi-clonal seeds. It has been reported that oil yields of UPB semi- and bi-clonal D×P seeds ranged from 7.95 to 9.52 t/ha/yr (Sharma 2006). However, Djunjana et al. (2011) and Zulhermana et al. (2011) reported that there was no statistical difference between semi-clonal and conventional D×P seeds.

2.6 Conclusion

The role of genetic resources is crucial for future developments since successful breeding and selection depend on a great deal of genetic variability. Several organizations have put up efforts to assemble *E. guineensis* and *E. oleifera* genetic materials from Africa and Central and South America. Introducing these materials into the present breeding populations will enable development of planting materials with higher oil yields, productivity and diversified products. Such endeavour will also ensure variability among oil palm commercial materials for oil palm sustainability.

References

Alwee SSRS, Tan JS, Nurul FFH, Izwanizam A, Yahya AK (2017a) Exploiting germplasm for oil palm trait development—a case for drought tolerance. Paper presented at Malaysian International Genetics Congress, Bangi-Putrajaya Hotel, Selangor, 25–27th September 2017

Alwee SSRS, Tan JS, Lee YP, Mohd Na'aim S, Haryati A, Leao LJ, Muhamad Farid R, Suthashinikisan K, Kwan YY (2017b) Five decades of oil palm breeding in FGV and moving forward. Paper presented at 13th ISP National Seminar 2017, Dorsett Grand Subang, Selangor, 17–19th July 2017

Arolu IW, Rafii MY, Marjuni M, Hanafi MM, Sulaiman Z, Rahim HA, Kolapo OK, Abidin MIZ, Amiruddin MD, Din AK, Nookiah R (2016) Genetic variability analysis and selection of pisifera palms for commercial production of high yielding and dwarf oil palm planting materials. Ind Crops Prod 90:135–141

Arulraj S, Suresh K (2011) Directorate of oil palm research vision 2030. Indian Council of Agricultural Research, Pedavegi, p 65

Ataga CD, Okwuangku CO, Okolo EC (1999) Characterisation of a recent oil palm (*Elaeis guineensis* Jacq.) germplasm collection and its exploitation in Nigeria. In: Preprints, PORIM international palm oil conferences, Kuala Lumpur, Malaysia, pp 277–280

Auxilia J, Shabha N (2017) Breeding of fruit and plantation crop. Agrimoon.com, p 220

Bakoumé C (2016) Genetic diversity, erosion, and conservation in oil palm (*Elaeis guineensis* Jacq.). In: Ahuja M, Jain S (eds) Genetic diversity and erosion in plants. Sustainable development and biodiversity. Springer, Cham, vol 8, pp 1–33

Beirneart A (1993) La selection du palmier a huile. Bulletin Agricole du Conggo Belge 2:359–380

Chin CW, Suhaimi S (1999) FELDA oil palm planting materials. In: Proceedings of the sourcing of oil palm planting materials for local and overseas joint ventures, pp 71–90

Cochard B, Amblard P, Durand-Gasselin T (2005) Oil palm genetic improvement and sustainable development. Oléagineux, Corps gras, Lipides 12:141–147

De Berchoux C, Gascon JP (1965). Caractéristiques végétatives de cinq descendances d'Elaeis guineensis Jacq. Premières données biométriques relations entre divers dcaractères et la production

Djunjana J, Hadi P, Joko H, Iswandar HE, Sitepu B, Nelson SP (2011) Performance of Sumbio semi-clonal progenesis. In: Proceeding of PIPOC 2011 international palm oil congress: palm oil fortifying the world, Kuala Lumpur Convention Centre, Kuala Lumpur, 15–17 November 2011

Escobar R (1982) Preliminary results of the collection and evaluation of the American oil palm Elaeis oleifera (HBK, [Humbaldt Bonbaldt, Kunith] Cortes) in Costa Rica [Panama and Colombia]. In: International conference on oil palm in agriculture in the eighties. Session-breeding, genetics and selection II, Kuala Lumpu, Malaysia, 17–20 June 1981

Hardon JJ, Tan GY (1969) Interspecific hybrids in the genus Elaeis I. crossability, cytogenetics and fertility of F 1 hybrids of *E. guineensis* × *E. oleifera*. Euphytica 18(3):372–379

Hardon JJ, Thomas RL (1968) Breeding and selection of the oil palm in Malaya. Oleagineaux 3:85–90

Hardon JJ, Corley RHV, Lee CH (1987) Breeding and selecting the oil palm. Improving vegetatively propagated crops. Academic Press, London, pp 63–81

Hartley CWS (1988) The oil palm (*Elaies guineensis*) Jacq.). Longman Scientific and Technical Publication, Wiley, New York. Tropical Agric. Series, Longman, p 761

Isa ZA, Mohd Din A, Maizura I, Noh A, Kushairi A, Rajanaidu N (2006) PS12: Breeding population for high oleic acid palm oil. MPOB Information Series No 313. MPOB, Bangi, Malaysia

Jagoe RB (1952) The 'dumpy' oil palm. Malaya Agric J 35:12

Kushairi A, Rajanaidu N (2000) Breeding populations, seed production and nursery management. In: Advances in Oil Palm Research, Malaysian Palm Oil Board, Bangi Selangor, Malaysia, vol 1, pp 39–96

Kushairi A, Rajanaidu N, Jalani BS (1998) The performance and genetical variation for agronomic traits in four progeny trials on coastal and inland soils. PORIM Viva Committee. Viva no. 0061(16). PORIM

Kushairi A, Rajanaidu N, Rafii MY, Mohd Din A, Mohd Isa ZA (2001) Population 12-Pollen for the development of dwarf PS1. MPOB Information Series No 15. Malaysian Palm Oil Board, Bangi, Malaysia

Kushairi A, Rajanaidu N, Mohd Din A, Isa ZA, Noh A, Junaidah J (2003a) PS5: breeding populations selected for thin shell Tenera. MPOB information series no 184. Malaysian Palm Oil Board, Bangi, Malaysia

Kushairi A, Rajanaidu N, Mohd Din A, Mohd Isa ZA, Noh A, Junaidah J (2003b) PS6: breeding populations selected for large fruit duras. MPOB Information Series No 183. Malaysian Palm Oil Board, Bangi

Kushairi A, Rajanaidu N, Sundram K, Maizura I (2004) PS8: breeding populations selected for high vitamin E. MPOB Information Series No 222. Malaysian Palm Oil Board, Bangi

Kushairi A, Tarmizi AH, Zamzuri I, Ong-Abdullah M, Samsul Kamal R, Ooi SE, Rajanaidu N (2010) Production, performance and advances in oil palm tissue culture. Paper presented at the ISOPB International Seminar on Advances in Oil Palm Tissue Culture, Yogyakarta, Indonesia, 29 May 2010

Maizura I, Kushairi A, Mohd Din A, Noh A, Marhalil M, Wong YT, Sambanthamurthi R (2008) PS13: breeding populations selected for low lipase. MPOB Information Series. MPOB Information Series. Malaysian Palm Oil Board, Bangi

Marhalil M (2009) Perfromance of MPOB-Tanzania Oil Palm Germplasm collection. MPOB 134th VIVA Committee Meeting. MPOB, Bangi.

Marhalil M, Rajanaidu N, Mohd Din A, Suzana M, Zulkifli Y, Fadila A, Saleh GB, Kushairi A (2016) Oil palm breeding for seed production in MPOB and introduction of nigerian-based D×P population. Paper presented at ISOPB Seminar, Kisaran, Indonesia, 29–30 September 2016

Marhalil M, Rajanaidu N, Mohd Din A, Zulkifli Y, Norziha A, Fadila AM, Suzana M, Ong MA, Saleh GB, Kushairi A (2017) Mining the exotic Nigerian Population 12. Paper presented at OralMPOB International Palm Oil Congress & Exhibition (PIPOC 2017). Kuala Lumpur Convention Centre, Kuala Lumpur, 14–16 November 2017

Meunier J (1975) The American oil palm Elaeis melanococca. Oleagineux 30:51–61

Mohd Din A (2000) Genetic variation for yield, bunch components and vegetative traits in oil palm (*Elaeis oleifera*) and interspecific hybrids. PhD thesis. Universiti Kebangsaan Malaysia

Mohd Din A, Rajanaidu N, Jalani BS (2000) Performance of *Elaeis oleifera* from Panama, Costa Rica, Honduras and Colombia in Malaysia. J Oil Palm Rest 12(1):71–80

Nampoothiri KUK, Pillai RSN (1998) Crop improvement in oil palm—present status and future strategies. In: Rethinam P, Suresh K (eds) Oil palm research and development. In: Proceedings of the national seminar on opportunities and challenges for the oil palm development in the twenty first century. Vijayawada, India, 19–21 January 1998

Noh A, Kushairi A, Mohd Din A, Maizura I, Isa ZA, Rajanaidu N (2005) PS10: breeding populations selected for long stalk. MPOB Information Series No 263. Malaysian Palm Oil Board, Bangi

Okoye MN, Bakoume C, Uguru MI, Singh R (2018) Genetic diversity and relatedness of oil palm progenies determined by microsatellite and agronomic markers. Afr J Biotech 17(18):614–625

Okwuagwu CO (1986) The genetic basis of the NIFOR oil palm breeding programme. In: Proceeding of international workshop on oil palm germplasm and utilization. Workshop Proceedings. Palm Oil Research Institute of Malaysia, pp 228–237

Ooi SC, Da Silva EB, Muller AA, Nascimento JC (1981) Oil palm genetic resources native *Elaeis oleifera* populations in Brazil offer promising sources. African oil Plant breeding; Cross-breeding; Hybrid progenie. Pesquisa Agropecuaria Brasileira (Brazil) 16(3):385–395

Pedapati A, Tyagi V, Yadav S, Brahmi P, Murugesan P (2013) Present status and future priorities for introduction of oil palm in India. The Ecoscan 3:134–144

Pillai RSN, Blaak G, Paul Closen H (2000) Collection of oil palm (*Elaeis guineensis* Jacq.) germplasm from Africa. Int J Oil Palm 1:23–37

Rajanaidu N (1983) Collection of oil palm (Elaeis oleifera) genetic material in Central and South America. PORIM Bull 6:1–11

Rajanaidu N, Ainul MM (2013) Conservation of oil palm and coconut genetic resources. In: Normah M, Chin H, Reed B (eds) Conservation of tropical plant species. Springer, New York, NY

Rajanaidu N, Rao V (1987) Oil palm genetic collections, their performance and use to the industry. In: Proceedings of the international oil/palm/palm oil conferences. Progress and prospects. (Agriculture) Kuala Lumpur (No. L-0115). PORIM

Rajanaidu N, Kushairi A, Rafii MY, Mohd Din A, Maizura I, Jailani BS (2000) Oil palm breeding and genetic resources. In: Yusof B, Jalani BS, Chan KW (eds) Advances in oil palm research, vol 1. Malaysian Palm Oil Board, pp 171–237

Rajanaidu N, Kushairi A, Marhalil M, Mohd Din A, Fadila AM, Noh A, Meilina A, Raviga S, Isa ZA (2013a) Breeding for oil palm for strategic requirement of the industry. In: Proceeding of MPOB International Palm Oil Congress (PIPOC 2013). Kuala Lumpur, pp 19–22

Rajanaidu N, Ainul MM, Kushairi A, Mohd Din A (2013b) Historical review of oil palm breeding for the past 50 years—Malaysian journey. In: Proceeding of the international seminar on oil palm breeding—yesterday, today and tomorrow, Kuala Lumpur, Malaysia, pp 11–28

Rajanaidu N, Kushairi A, Mohd Din A (2017) Monograph oil palm genetic resources. Malaysian Palm Oil Board, p 289

Rosenquist EA (1986) The genetic base of oil palm breeding populations. In: Procceding of international workshop on oil palm Germplasm and utilization, Bangi, Selangor-Malaysia (No. L-0177). PORIM, 26–27 March 1986

Sharma M (1999) Utilization of Nigerian PS1 and PS2 selectionin oil palm breeding programmes at UP BHD. In: Rajanaidu N, Jalani BS (eds) Proceedings of seminar on PS1 and PS2 oil palm planting materials. Palm Oil Research Institute of Malaysia, Bandar Baru Bangi, Selangor, pp 18–29

Sharma M (2006) Challenges facing the Malaysian palm oil industry-multi pronged strategies for raising oil yield, productivity and profitability. In: Kushairi A, Sambanthamurthi R, Ong Abdullah M, Kwong Choong C (eds) Proceeding of clonal & quality. Replanting material workshop. Malaysian Palm Oil Board, Kuala Lumpur, pp 56–94

Singh R, Ong-Abdullah M, Low ETL, Manaf MAA, Rosli R, Nookiah R, Ooi LCL, Ooi SE, Chan KL, Halim MA, Azizi N (2013) Oil palm genome sequence reveals divergence of interfertile species in old and new worlds. Nature 500(7462):335

Soh AC,Wong CK, Tan CC, Wong G, Hor TY (2005) Super yielding oil palms—strategic breeding plan. Paper Presented at National Seminar on advances in breeding and clonal technologies for super yielding planting materials. Malaysian Palm Oil Board, Bangi

Sparnaaij LD, Menendez T, Blaak G (1963) Breeding and inheritance in the oil palm oil (*Elaies guinensis Jacq.*) Part 1: The design of a breeding programme. J West Afr Inst Oil Palm Res 4:126–155

Suprianto E, Yenni Y, Supena N, Arif M, Sujadi, Siregar HA, Rahmadi HY, Wening S, Setiowati RD, Faizah R, Purba AR (2016) Current Status of IOPRI oil palm breeding programme and seed production. Paper presented at ISOPB Seminar. Kisaran, Indonesia, 29–30 September 2016

Vanderweyen R (1952) La prospection des palmeraies congolaises et ses premier results. Bull Inf INEAC 1:357–382

Veerappan P, Bilal M, Nazeeb M, Loong SG (2000) Early performance and potential of semiclonal D×P progenies. In: Pushparajah E (ed) Proceedings of international planters conference on plantation tree crops in the new millennium: the way ahead, vol I. The Incorporated Society of Planters, Kuala Lumpur: pp 117–130

Zeven AC (1967) The semi-wild oil palm and its industry in Africa. Wageningen Centre Agric. Publishing documentation

Zulhermana, Yulia P, Mario RS, Ario PR, Asmady, Dwi A (2011) Current progress on Sampoerna Agro D×P semi-clones. In: Proceeding of international palm oil congress (PIPOC), Kuala Lumpur Convention Centre, 15–17 November 2011, pp 266–270

Breeding and Improvement of the Oil Palm Interspecific Hybrids via Molecular Cytogenetics

3

M. Madon, H. M. N. Nordiana, Z. Muhammad Azwan, and N. M. Zaki

Abstract

Breeding and improvement of the oil palm interspecific hybrids can take long time and involve multiple resources due to oil palm having a long breeding cycle, which needs large areas to be planted, intensive labour, and high maintenance cost. However, efforts can be optimized via molecular cytogenetics such as fluorescence in situ hybridization (FISH) method to locate DNA sequences of interest on chromosomes and genomic in situ hybridization (GISH) which is used to discriminate chromosomes of *E. guineensis* and *E. oleifera* by ascertaining the amount of oleifera genomes introgressed into O×G hybrids and their backcrosses and to look for introgressed chromosomes. The oil palm planted area worldwide is increasing continuously. Therefore, it is necessary to be prepared with a diversified planting material to avoid major economic loss if ever the disease arises in Malaysia.

3.1 Introduction

Malaysia is blessed with the oil palm, *E. guineensis* which was brought in by Henry Fauconnier in the year 1917 (Kushairi et al. 2017). It is widely cultivated in the tropical countries where adequate rainfall, sufficient sunshine, and optimal soil conditions such as in Southeast Asia, particularly Malaysia and Indonesia, are suitable for oil palm to grow (Kushairi et al. 2017). The oil palm produces palm oil and palm kernel oil which can be used for both food and non-food uses, and it is the most productive and yields four to ten times higher than any oil crop (Soh et al. 2017). Up till 2018, the total area in Malaysia planted with oil palm *E. guineensis* fruit type *tenera* (D×P) is 5.81 million hectares (Kushairi et al. 2018).

This is a cause of concern as we should learn from history of the devastating Irish potato famine in 1845. During this period, HERB-1, a strain of *Phytophthorainfestans* has caused blights to the potato Irish lumper planted extensively as a monocrop. By late 1845, the year's harvest was destroyed, and the first deaths due to starvation were reported. For oil palm, in the case of bud rot, the first epidemic reaching catastrophic proportion was reported in Colombia in the year 1964. However, recent epidemics destroyed more than 70,000 ha in the western and central oil palm growing regions of Colombia. This is a destructive emerging disease of oil

M. Madon · H. M. N. Nordiana · Z. Muhammad Azwan · N. M. Zaki (✉)
Malaysian Palm Oil Board, No. 6, Persiaran Institusi, Bandar Baru Bangi, 43000 Kajang, Selangor, Malaysia
e-mail: hariza@mpob.gov.my

© Springer Nature Switzerland AG 2020
M. Ithnin and A. Kushairi (eds.), *The Oil Palm Genome*, Compendium of Plant Genomes, https://doi.org/10.1007/978-3-030-22549-0_3

palm, and the alternative was to replace *E. guineensis* with O×G hybrids due to *E. oleifera* exhibiting partial tolerance to bud rot (Torres et al. 2016). The bud rot is caused by *Phytophtrapalmivora,* and in the tropics, *P. palmivora* is a pathogen of many plant species. Due to the extensive monocropping of *tenera* in Malaysia, hence, it is felt that breeding of interspecific hybrids need to be developed as a backup for managing risk of bud rot outbreak.

3.2 *Elaeis guineensis, Elaeis oleifera* and Interspecific Hybrids

Elaeis guineensis Jacq. (G) and *Elaeis oleifera* (HBK) cortès (O) are the two species of oil palm. Both species have $2n = 32$ chromosomes which can be divided into group I (pair no. 1), group II (pairs no. 2–9) and group III (pairs no. 10–16) (Madon et al. 1998). Figure 3.1a, b show the karyo types of *E. guineensis* and *E. oleifera,* respectively. The centromeres of *E. guineensis* chromosomes are less significant compared to *E. ole*ifera, and hence, *E. oleifera* chromosomes can be distinguished into submetacentrics, subtelocentrics and telocentrics. Recently, Singh et al. (2013) have re-assigned the 16 pairs of *E. guineensis* individual chromosomes into four groups: Group I, the largest chromosomes (hybridizes to 5S rDNA at the proximal); Group II, 8 medium chromosomes; Group III, 6 small chromosomes; and Group IV, small acrocentric chromosomes (hybridized to 18S–25S rDNA). For *E. guineensis*, the leaflet plane is double, while in *E. oleifera* (Fig. 3.2a, b) it is single. The shapes of the microspores or immature pollens of *E. guineensis*, *E. oleifera* and O×G hybrids are triangular, oblong and a combination of a triangular base and an oblong tip, respectively, as shown in Fig. 3.3a–c.

Elaeis guineensis produces high oil yields while *E. oleifera* produces more unsaturated oil with high carotene content, short in stature, and has lower lipase activity, higher vitamins A and E contents and high levels of pest and disease tolerance (Barcelos and Amblard 2017). However, *E. oleifera* produces lower oil yield; hence, the two species are crossed to produce O×G hybrids, and when compared to *E. guineensis*, the O×G hybrids have more unsaturated oil, lower height increment, oil with high carotene content and exhibit disease tolerance. In Latin America, the F1 O×G hybrids are widely planted due to disease tolerance and the increasing efforts to introgress the interesting *E. oleifera* traits into high yielding *E. guineensis* varieties (Barcelos and Amblard 2017). However, the hybrids suffer from low oil yield, vigorous vegetative growth and low natural fertility which requires assisted pollination, and hence, incurring high costs and labour. Hardon and Tan (1969) also found that in O×G hybrids, the bunch set, seedling germination and survival were all well below that of *E. guineensis*. While Schwendimen et al. (1983) also reported that due to a poorly understood cytogenetics problem, the OG hybrids have some reproductive limitations. To improve these characteristics, O×G hybrids are backcrossed to their *guineensis* parents. Obasola et al. (1976)

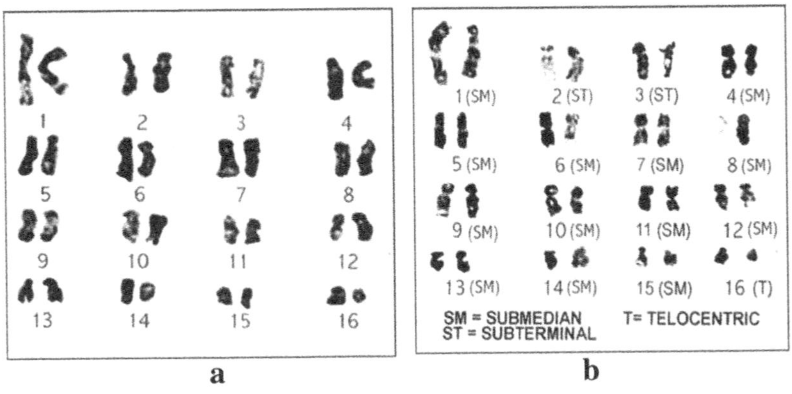

Fig. 3.1 Karyotypes of *E. guineensis* and *E. oleifera* chromosomes

Fig. 3.2 Horizontal view showing **a** *E. oleifera* with single plane leaflet arrangement and **b** *E. guineensis* with double plane leaflet arrangement

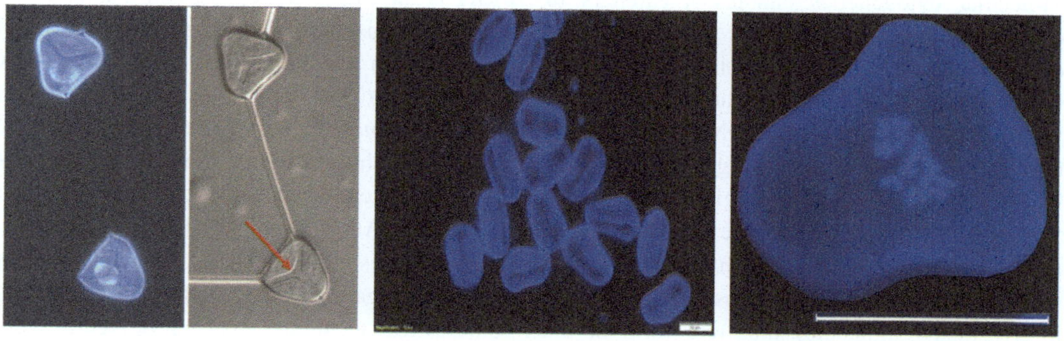

Fig. 3.3 Immature pollens or microspores morphology are **a** triangular for *E. guineensis*, **b** oblong for *E. oleifera* and **c** combination of triangular base and oblong tip for O×G hybrids

conducted a study on a backcross programme of O×G hybrid with *E. guineensis* (*tenera*) as the male parent, and it showed that the progenies exhibited three morphological categories. They are like E. *guineensis*, *E. oleifera* and those combining characters of both species.

Malaysian Palm Oil Board (MPOB) has carried out extensive germplasm collections in order to broaden the genetic base of current breeding materials which originated from the four Deli *dura* palms planted in Bogor Botanical Garden in 1848 (Hartley 1988). The germplasm collections were done in Africa and Latin America. For *E. guineensis*, the germplasms collected were from 11 countries, namely Nigeria, Cameroon, Zaire, Tanzania, Madagascar, Angola, Senegal, Gambia, Sierra Leone, Guinea and Ghana (Rajanaidu et al. 2017). While for *E. oleifera*, the collections came from seven countries, and they are Honduras, Panama, Costa Rica, Suriname, Ecuador, Peru and Columbia (Rajanaidu et al. 2017). These germplasms provide new genetic materials that harness potential new traits of economic interest. These traits then can be discovered and subsequently introgressed into the current breeding materials, and hence, broadening the genetic base (Mayes et al. 2017). Using the *E. guineensis* and *E. oleifera* germplasms, O×G hybrids and backcrosses can be created to obtain good combinations of interesting traits. However, these efforts can take a long time and involve multiple resources due to oil palm having a long breeding cycle, which needs large areas to be planted, intensive labour and high maintenance cost. Biotechnology offers a way forward for oil palm interspecific hybrid breeding and improvement programmes particularly molecular cytogenetics which combines molecular biology and cytogenetics.

3.3 Molecular Cytogenetics

Molecular cytogenetics enables researchers to observe sequence of interest or DNA fragments on the physical chromosomes themselves. This can be done via in situ hybridization experiments which is a technique to locate DNA sequences on chromosomes. It involves using a mixture containing segments of labelled single-stranded DNA which are incubated in suitable conditions with metaphase chromosomes which have also been rendered single stranded on glass slides, resulting in hybridized molecules that will form at the complementary bases. Previously, in situ hybridization used radioactive-labelled probes (Harper et al. 1981). However, the radioactive in situ hybridization technique offers limited resolution, requires long exposure time and needs appropriate procedures for radioactive waste disposal, and finally, the slides cannot be kept for a long period due to radioactive decay period.

Due to the above reasons, non-radioactive techniques, namely non-fluorescent and fluorescent techniques, were developed by Langer-Safer et al. (1982). Rayburn and Gill (1985) first introduced ISH technique in plant species where the probe DNA was labelled with biotinylated dUTP, and the hybridization sites were detected with horseradish peroxidase or alkaline phosphatase conjugated avidin or streptavidin (Jiang and Gill 1994). The hybridization sites appear as banding pattern due to chromogenic substrate, and the slides can be kept for a long time. However, now, researchers have moved to using fluorescence in situ hybridization (FISH) due to its flexibility and versatility. It has a wide variety of applications, relatively easy to be implemented and provides high performance of in situ studies (Levsky and Singer 2003). The basic principles of FISH have remained unchanged, but high-sensitivity detection, simultaneous assay of multiple species and automated data collection and analysis have advanced the field significantly and have become a foremost biological assay.

3.4 Fluorescence in Situ Hybridization (FISH)

FISH is an extremely useful addition to study plant chromosomes. The application of FISH technique on oil palm was first reported by Madon et al. (1996) using rDNA probe from flax to *Elaeis guineensis* (*tenera*) chromosomes. The rDNA probe hybridized on the telomeric regions of acrocentric chromosome pair and on their satellite DNAs. This is the shortest pair of oil palm chromosomes which is also referred to as nucleolar organizer ribosomal (NOR) chromosomes. FISH technique has been proven to work on oil palm chromosomes, therefore enabling other DNA sequences to be located as reported by Castilho et al. (2000). The sequences localized were 5S rDNA, 18S-25S rDNA, *copia* retrotransposons and repetitive DNA clones. This information provides some insight toward understanding the oil palm genome. Singh et al. (2013) reported the 1.8–gigabase (Gb) genome sequence of the African oil palm *Elaeis guineensis*, the predominant source of worldwide commercial planting for oil production and the draft sequence of the south American oil palm *E. oleifera*, which has the same number of chromosomes ($2n = 32$) and produces fertile interspecific hybrids with *E. guineensis* but yet seems to be diverged in the new World. The genome sequence also enables the discovery of genes for important traits.

Zaki et al. (2017) have utilized the oil palm genome sequence in order to develop the *E. guineensis* chromosome-arm specific markers that can link the genetic, sequence and chromosomal maps, and hence, valuable for comparative, hybridization and breeding studies. It was found that the conserved putative repetitive DNA sequence found through informatics analysis showed an extra intercalary band in one arm of pseudo-chromosome 1 in FISH analysis, thus establishing the north-south orientation of the *E. guineensis* pseudo chromosome. The newly developed marker is also able to distinguish both *Elaeis* species with FISH indicating its utility in identifying *Elaeis* hybrids (Zaki et al. 2017).

Marker-assisted breeding is an important tool for genetic improvement programme of species that has extremely long selection cycle like oil palm (Ting et al. 2014). Various molecular markers such as AFLP, RFLP and SSR were used to develop genetic linkage maps and QTLs for fatty acid composition, and yield components were identified (Rance et al. 2001; Billotte et al. 2010; Singh et al. 2009). However, the markers detected from low density genetic map are less accurate, and hence, their application is limited. Ting et al. (2014) have genotyped SNP and SSR markers onto intraspecific and interspecific oil palm hybrids, and this has resulted in high-density SNP and SSR-based genetic maps. The marker density and genome coverage were greatly improved compared to the first reference map based on AFLP and SSR markers (Ting et al. 2014). The markers detected from high-density genetic map will in future provide useful and interesting markers that can be located using FISH on the oil palm chromosomes which will assist in the selection of planting material.

3.5 Genomic in Situ Hybridization (GISH)

GISH technique utilizes both genomic and blocking DNA while FISH only uses DNA probes that vary in length without any block (Ramzan et al. 2017). GISH is a tool to analyse genomic structure and function, chromosome constituents, recombination patterns, alien gene introgression, genome evolution, aneuploidy and polyploidy, genome constitution visualization and chromosome discrimination from different genomes in allopolyploids (Ramzan et al. 2017). GISH was first developed by Schwarzacher et al. (1989) to characterize the parental genomes of the intergeneric hybrid *Hordeumchilense* and *Secale africanum* where labelled total genomic DNA of a plant species is used as probe on the chromosome spreads on intergeneric hybrids. It provides an effective way for parental genome analysis in both sexual and somatic interspecific hybrids and enables identification of intergenomic recombination chromosomes from both parental species, and their behaviour and transmission also can be monitored through meiosis into the progeny of subsequent backcrosses (Parokonny et al. 1997). Ramzan et al. (2017) reviewed the application of GISH in horticultural plants recently and concluded that chromosomal evaluation, cytogenetical classification, genomic constitution, polyploidy confirmation, hybrid verification and introgression breeding have been performed using GISH.

Improvement of interspecific hybrids is intended to introgress desirable traits from *oleifera* into *guineensis*. GISH can be done on both mitotic and meiotic chromosomes. In oil palm, the mitotic chromosomes are obtained from the root tip meristematic nuclei while meiotic chromosomes from the pollen mother cells in the anthers of male flowers. For interspecific F1 O×G hybrids mitotic chromosomes, Madon et al. (1999) have utilized the genomic DNA from *E. oleifera* as probe by labelling it with biotin while 90 µg of *E. guineensis* genomic DNA was used as block. The hybridization sites were detected with FITC-avidin, and the results showed that the F_1 O×G hybrids contained 16 *E. oleifera* and 16 *E. guineensis* chromosomes (Fig. 3.4a, b). Previously, it was assumed that the F1 O×G hybrids contain 16 *E. oleifera* and 16 *E. guineensis* chromosomes, and with this GISH technique, the assumption is proven right. In order to further improve the O×G hybrids, they were backrossed

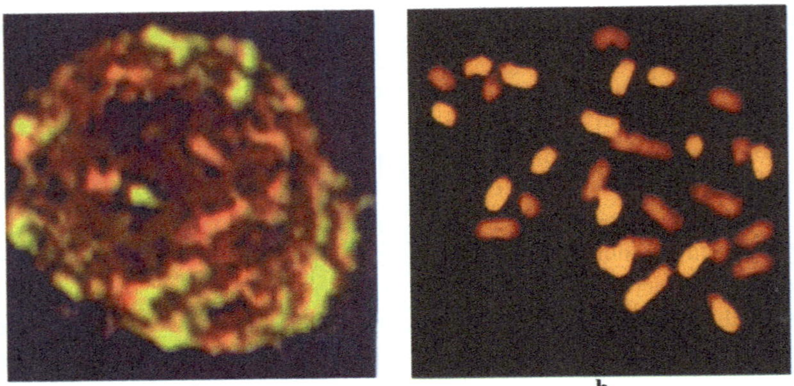

a b

Fig. 3.4 GISH of O×G hybrid probed with biotin labelled total genomic DNA from *E. oleifera* and blocked with genomic DNA from *E. guineensis*. **a** An interphase nucleus of O×G showing groups of chromosomes from both genomes occupying, discrete, non-intermixed domains. *E. oleifera* genome fluoresced yellow and *E. guineensis* genome fluoresced red. **b** Metaphase spread of above showing 16 *E. oleifera* (yellow) and 16 *E. guineensis* (red) chromosomes (Reproduced from Madon et al. 1999)

to the *guineensis* parent, and it was assumed that the genetic contribution of *E. oleifera* will be reduced by an average of 50% in each successive backcrossing to the *guineensis* parent, and each backcross cycle requires at least 10–12 years for the progenies to mature. Madon et al. (2018) has utilized GISH technique on a few backcross one progenies of O×G crossed to *E. guineensis*. Both species share high genome homology, and hence, 120 µg of unlabelled *E. guineensis* DNA was used as block to increase specificity of probing the backcross one chromosomes. The blocking DNA hybridized to the sequences common in the blocking DNA, probe DNA and chromosomal DNA in situ, thereby leaving mainly species specific sequences as sites for labelled probe hybridization (Orgaard and Heslop-Harrison 1994). GISH experiments on the backcrosses showed the presence of variable numbers of *E. oleifera* chromosomes (Fig. 3.5a1–c1). Hence, proving the assumption that the genetic contribution of *E. oleifera* will be reduced by an average of 50% in each successive backcrossing to the *guineensis* parent is inaccurate.

During meiosis in backcross hybrids, each gamete receives variable composition of *E. oleifera* and *E. guineensis* chromosomes while maintaining $2n = 32$. In meiosis, there are several processes such as independent assortment, random pairing and crossing over which determine the backcross individuals genetic make-up, and it can consist of either *E. guineensis* genome ($2n = 32$) or a variable number of *E. oleifera* chromosomes, with maximum number of 16 out of 32 chromosomes. The chromosome composition in the backcross palms showed that they are mostly composed of Groups II and III chromosomes with $2n = 32$. Chromosomes originating from the two different species occupy discrete, non-intermixed domains in interphase nuclei of O×G hybrids indicating the non-random organization of interphase nucleus (Madon et al. 1999). Similar observations were also seen in the interphase nuclei of BC1 (Fig. 3.5a2–c2) (Madon et al. 2018). Ideally, GISH can be used to discriminate chromosomes of both species and ascertain the amount of *oleifera* genomes introgressed into O×G hybrids and their backcrosses and to look for introgressed chromosomes. However, due to the small size of oil palm chromosomes, recombined chromosomes were difficult to observe. In plant breeding programmes in which the genetic base of a crop can be expanded by chromosome transfer from

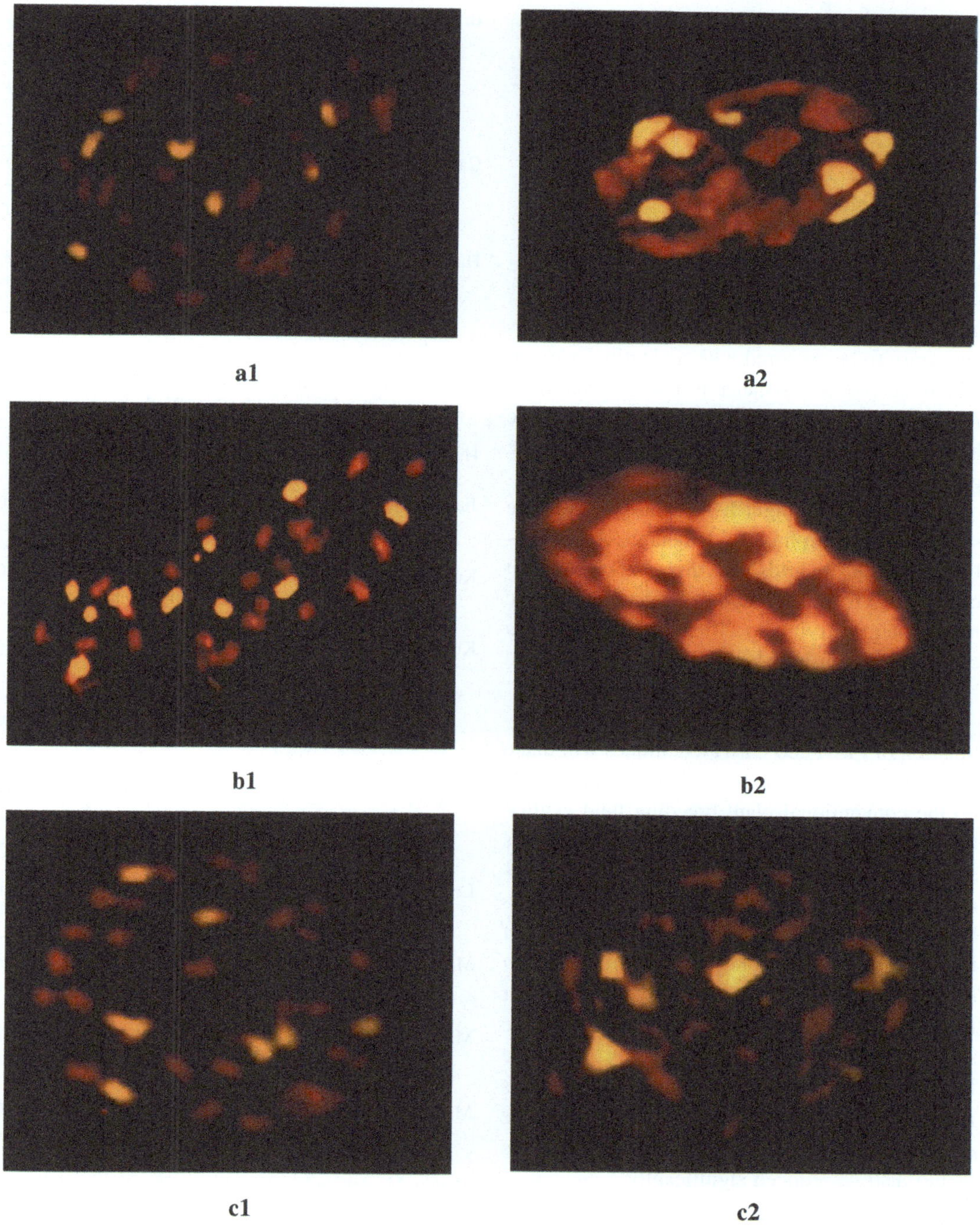

Fig. 3.5 **a**1 GISH of BC1 palm A showed 7 O (yellow) and 25 G (red), **b**1 BC1 palm B showed 10 O (yellow) and 22 G (red) and **c**1 BC1 palm C showed 8 O (yellow) and 24 G (red) chromosomes while (**a**2, **b**2 and **c**2) showed groups of chromosomes from both genomes occupying, discrete, non-intermixed domains in interphase nuclei (Reproduced from Madon et al. 2018)

wild or another species of the same genus, non-homologous (intergenomic) recombination, somatic or meiotic, is desirable (Anamthawat-Jonsson and Bodvarsdottir 1998).

3.6 Conclusion

Hybrid breeding is important for combining the good traits present in *E. oleifera* into *E. guineensis* elite oil palm planting material. It is necessary to be prepared with a planting material which has partial resistance to bud rot caused by *Phytopterapalmivora* to avoid major economic loss if ever the disease arises in Malaysia. The genomic in situ hybridization technique developed is able to distinguish parental chromosome sets in both interphase and metaphase preparations of O×G hybrids and their backcrosses. It can be used by the breeders to select interspecific hybrid palm carrying specific number of *oleifera* chromosomes in breeding populations generated by conventional plant breeding programme.

With the advent of molecular markers, combination of FISH and GISH techniques will make it possible for selection to be done earlier, speeds up the conventional plant breeding field evaluation and reduces the number of individuals selected for further backcrossing. Hence, superior backcross palms containing desirable numbers of *E. oleifera* chromosomes with markers of interest or presence of *guineensis* or *oleifera* recombinant chromosomes can be selected for further backcrossing programmes. Since the screening can be done at the nursery stage, the land area required to plant the selected backcross materials will be reduced. In so doing, the duration of backcross breeding programmes would also be reduced significantly.

References

Anamthawat-Jonsson K, Bodvarsdottir SK (1998) Meiosis of wheat × lymegrass hybrids. Chromosome Res 6:339–343

Barcelos E, Amblard P (2017) *Elaeisoleifera* genetic resources. In: Soh AC, Mayes S, Roberts J (eds) Oil palm breeding. Genetics and genomics, pp 45–48

Billotte N, Jourjon MF, Marseillac N, Berger A, Flori A, Asmady H, Adon B, Singh R, Nouy B, Potier F, Cheah SC, Rohde W, Ritter E, Courtois B, Charrier A, Margin B (2010) QTL detection by multi-parent linkage mapping in oil palm (*Elaeis guineensis* Jacq.). Theor Appl Genet 120(8):1673–1687

Castilho A, Vershinin AV, Heslop-Harrison JS (2000) Repetitive DNA and the chromosomes in the genome of oil palm (*Elaeis guineensis*). Ann Bot 856:837–844

Hardon JJ, Tan GY (1969) Interspecific hybrids in the genus *Elaeis* I. Crossability, cytogenetics and fertility of F1 hybrids of *E. guineensis* × *E. oleifera*. Euphytica 18:372–379

Harper ME, Ulrich A, Saunders GF (1981) Localization of the human insulin gene to the distal end of the short arm of chromosome 11. Proc Natl Acad Sci USA 78:4458–4460

Hartley CWS (1988) The oil palm (*Elaeis guineensis* Jacq.) Longman Wiley, New York

Jiang J, Gill BS (1994) Non-isotopic *in situ* hybridization and plant genome mapping: the first 10 years. Genome 37:717–725

Kushairi A, Singh R, Ong-Abdullah M (2017) The oil palm industry in Malaysia: thriving with transformative technologies. J Oil Palm Res 29(4):431–439

Kushairi A, Soh KL, Azman I, Hishamuddin Elina, Ong-Abdullah M, Izuddin Zainal BidinMohd Noor, Razmah G, ShamalaSundram Ghulam Kadir, Parveez Ahmad (2018) Oil palm economic performance in Malaysia and R&D progress in 2017. J Oil Palm Res 30(2):163–195

Langer-Safer PR, Levine M, Ward DC (1982) Immunological method for mapping genes on Drosophila polytene chromosomes. Proc Natl Acad Sci USA 79:4381–4385

Levsky JM, Singer RH (2003) Fluorescence in situ hybridization: past, present and future. J Cell Sci 116:2833–2838

Madon M, Clyde MM, Cheah SC (1996) Fluorescence in situ hybridization of rRNA probe to Elaeis guineensis (*tenera*) chromosomes. Elaeis 8(1):29–36

Madon M, Clyde MM, Cheah SC (1998) Cytological analysis of *Elaeis guineensis* and *Elaeis oleifera* chromosomes. J Oil Palm Res 10(1):68–91

Madon M, Clyde MM, Cheah SC (1999) Application of genomic *in situ* hybridization (GISH) on *Elaeis* hybrids. J Oil Palm Res (Spec Issue): 74–80

Madon M, Arulandoo X, Sritharan K, Nordiana HMN, Muhammad Azwan Z, Mohd Zaki N (2018) Short communication: genomic constitution of oil palm interspecific hybrid crosses monitored by genomic in situ hybridization (GISH). J Oil Palm Res (accepted)

Mayes S, Soh AC, Roberts J (2017) Genetic resources. introductory overview. In: Soh AC, Mayes S, Roberts J (eds) Oil palm breeding. Genetics and genomics, pp 21–23

Obasola CO, Obesesan IO, Opute FI (1976) Breeding of short-stemmed oil palm in Nigeria. In: Proceedings of international agriculture oil palm conference, Kuala Lumpur, 26–28 May

Orgaard M, Heslop-Harrison JS (1994) Investigations of genome relationships between *Leymus, Psathyrostachys* and *Hordeum* inferred by genomic DNA: DNA in situ hybridization. Ann Bot 73:195–203

Parokonny AS, Marshall JA, Bennett MD, Cocking EC, Davey MR, Power JB (1997) Homoeologous pairing and recombination in backcross derivatives of tomato somatic hybrids [*Lycopersiconesculentum* (+) *L. peruvianum*]. Theor Appl Genet 94:713–723

Rajanaidu N, Mohd Din A, Marhalil M, Abdullah N, Ong-Abdullah M, Ahmad Malike F, Abu Bakar NA, Libin A, Yaakub Z, Mustafa S, Ithnin M, Kushairi A (2017) Prospection, conservation and the broadening of the genetic base in oil palm. In: Soh AC, Mayes S, Roberts J (eds) Oil palm breeding. Genetics and genomics, pp 27–45

Ramzan, F, Younis A, Lim K-B (2017) Application of genomic in situ hybridization in horticultural science. Int J Genomics Article ID 7561909, 12 p

Rance KA, Mayes S, Price Z, Zack PL, Corley RHV (2001) Quantitative trait loci for yield components in oil palm (*Elaeis guineensis* Jacq.). Theor Appl Genet 103:1302–1310

Rayburn AL, Gill BS (1985) Use of biotin-labelled probes to map specific DNA sequences on wheat chromosomes. Heredity 76:78–81

Soh AC, Mayes S, Roberts J (2017) Introduction to the oil palm crop. In: Soh AC, Mayes S, Roberts J (eds) Oil palm breeding. Genetics and genomics, pp 1–6

Schwarzacher T, Leitch AR, Bennett MD, Heslop-Harrison JS (1989) In situ localization of parental genomes in a wide hybrid. Ann Bot 64:315–324

Schwendimen J, Pallares P, Amblard P, Baudouin L (1983) Analysis of fertility during bunch development in the interspecific oil palm hybrid *Elaeismelanococca* × *E. guineensis*. Oléagineux 38(7):411–420

Singh R, Tan SG, Panandam GM, Rahimah AR, Ooi LCL, Low ETL, Sharma M, Jansen J, Cheah SC (2009) Mapping quantitative trait loci (QTLs) for fatty acid composition in an interspecific cross of oil palm. BMC Plant Biol 9:114

Singh R, Ong-Abdullah M, Low ETL, Manaf MAA, Rosli R, Nookiah R, Ooi LCL, Ooi SE, Chan KL, Halim MA, Azizi N, Nagappan J, Bacher B, Lakey N, Smith SW, He D, Hogan M, Budiman MA, Lee EK, Desalle R, Kudrna D, Goicoechea JL, Wing RA, Wilson RK, Fulton RS, Ordway JM, Martienssen RA, Sambanthamurthi R (2013) Oil palm genome sequence reveals divergence of interfertile species in old and new worlds. Nature 500(7462):335–339

Ting NC, Jansen J, Mayes S, Massawe F, Sambanthamurthi R, Ooi LCL, Chin CW, Arulandoo X, Seng TY, Syed Alwee SSR, Ithnin M, Singh R (2014) High density SNP and SSR-based genetic maps of two independent oil palm hybrids. BMC Genom 15:309–320

Torres GA, Sarria GA, Martinez G, Varon F, Drenth A, Guest DI (2016) Bud rot caused by *Phytophthora palmivora*: a destructive emerging disease of oil palm. Phytopathology 106:320–329

Zaki MN, Singh R, Nordiana HMN, Zulkifli MA, Steven WS, Schwarzacher T, Madon M, Heslop-Harrison JS (2017) Short communication: towards development of *Elaeis guineensis* chromosome-arm specific markers and their utility across the *Elaeis* genus. J Oil Palm Res 29(4):594–599

Genetic Improvement of Oil Palm Through Recurrent Selection

4

Mohd Rafii Yusop, Jalani Sukaimi,
Mohd Din Amiruddin, Momodu Jalloh,
Senesie Swaray, Oladosu Yusuff, and
Samuel Chibuike Chukwu

Abstract

Plant breeding, particularly oil palm, takes up to 20 years per cycle for phenotypic selection which is characterized by low planting densities and longer generations. This, however, has numerous consequences such as limited breeding intensity and reduced number of generations that can be reached in between two breeding stages. In the attempt to reconcile the rapid achievement of varietal outputs and the long-term improvement of the tropical tree crops programme, the recurrent selection schemes were adopted. This paper presents an overview of recurrent selection with emphasis on modified reciprocal recurrent selection which is widely used in oil palm. Based on this background, the objectives of this breeding method were highlighted, and some achievements recorded through the method were elaborated.

4.1 Introduction

The natural populations of cross-pollinated plants are highly heterogeneous and heterozygous, i.e. they show both genotypic and phenotypic levels of variation. Different loci of these populations are expected to be at Hardy–Weinberg equilibrium in the absence of evolutionary influences such as migration, selection, genetic drift and mutation. Selection in these populations changes the genotype and gene frequencies along with the direction of selection. However, it is not likely to either fix or eliminate an allele of a gene. Furthermore, such populations will express variable degrees of inbreeding depression that may be rather severe in some plant species. Hence, it is highly necessary to develop varieties or breeding populations that are heterozygous so that their performance is not reduced due to inbreeding depression. Based on this background, the breeding schemes used for crop improvement

M. R. Yusop (✉) · M. Jalloh · S. Swaray · O. Yusuff · S. C. Chukwu
Institute of Tropical Agriculture and Food Security, Universiti Putra Malaysia, 43400 Serdang, Selangor, Malaysia
e-mail: mrafii@upm.edu.my

J. Sukaimi
Faculty of Science and Technology, Universiti Sains Islam Malaysia, 71800 Bandar Baru Nilai, Negeri Sembilan, Malaysia

M. D. Amiruddin
Malaysian Palm Oil Board, No. 6, Persiaran Institusi, Bandar Baru Bangi, 43000 Kajang, Selangor, Malaysia

M. R. Yusop
Department of Crop Science, Faculty of Agriculture, Universiti Putra Malaysia, 43400 Serdang, Selangor, Malaysia

S. C. Chukwu
Department of Crop Production and Landscape Management, Ebonyi State University, Abakaliki, Nigeria

© Springer Nature Switzerland AG 2020
M. Ithnin and A. Kushairi (eds.), *The Oil Palm Genome*,
Compendium of Plant Genomes, https://doi.org/10.1007/978-3-030-22549-0_4

have been premeditated to ensure that their end products such as lines/populations used as varieties are heterozygous and also free from inbreeding depression. In this chapter, breeding schemes used for cross-pollinated crops improvement are discussed in detail with major emphasis on recurrent selection. These breeding schemes can be classified into two broad groups, viz. (i) population or selection improvement and (ii) synthetic and hybrid varieties.

The breeding methods for cross-pollinated species were generally developed for maize, but these methods have subsequently been used for breeding of other cross-pollinated crop species. These breeding schemes have been designed to develop open-pollinated, synthetic, hybrid and composite varieties of cross-pollinated crop plants. Quite a number of breeding schemes applied for maize breeding such as topcross testing and recurrent selection have been adopted in oil palm. Moreover, being a perennial tree crop, it shares many selection techniques similar with animal breeding such as index or simultaneous multiple trait selection. For a better understanding of breeding methods designed towards sustainability of oil palm breeding, it is prerequisite to master the plant reproductive biology along with developments in modern and traditional breeding methods and innovative crop production concepts. For example, the traditional improvement scheme such as reciprocal recurrent selection for oil palm entails the use of different breeding methods to generate improved varieties or superior lines. However, these methods are expensive, and it takes an approximate of ten years of breeding cycle which comprise one year for pollination, two to three months for seed preparation and germination, one year in the nursery, three years in the field before first harvest and four to six years for yield evaluation (Kushairi et al. 2011; Rajanaidu et al. 2000). Adding these to progeny testing, it takes more than 20 years to develop a new planting material (Kushairi et al. 2011).

The objectives of selection in cross-pollinated crops can be summarized as follows: (i) increasing the frequency of desirable alleles in the selected population, (ii) increase in the frequency of the desirable genotypes, (iii) improving the characteristics of the population, such as the performance, (iv) maintaining the level of inbreeding to the minimum and (v) fixation of alleles to maintain homogeneity. The main objective of selection methods in cross-pollinated crops is to improve their relevant characteristics, including yield. Hence, the major objective in recurrent selection is to increase the favourable allele frequency that is affecting the trait of interest in order to increase the value of the population. Increasing the favourable gene frequency is advantageous for inbreeding to produce improved homozygous lines or population performance per se, as in the case of synthetic cultivars. The theoretical advantage of increased gene frequency prior to selection has been reported by Hallauer and Darrah (1985). The recurrent selection consists of repeated cycles of recombination and selection. Four major steps in recurrent selection include selection of the base population, progenies development, progenies evaluation and recombination of selected individuals.

4.2 Recurrent Selection

Plant breeding, particularly in tree crop, is a long-term programme. The time factor affects the breeding intensity and number of generations that could be accomplished in between two stages of breeding. Although the small number of generations per unit time reduces the risk of losing variation through genetic drift, the use of observations spanning through several years ensures more effectiveness in breeding strategies by reducing the effect of climatic hazards. The attempt to reconcile the rapid achievement of varietal outputs and the long-term improvement of the tropical tree crops programme has led to adoption of recurrent selection schemes.

The recurrent selection method is a modified progeny selection. The idea was first floated around 1920 which was later developed as a breeding scheme in 1945 when Hull suggested the technique for recurrent selection for specific combining ability (SCA). Basically, this technique was developed to facilitate heterosis breeding and not for populations improvement. The rationale behind this selection method is as follows. Inbreds used for the development of heterosis hybrid varieties are isolated from genetically variable, highly heterozygous open-pollinated populations known as base populations. Inbreds are homozygous lines of cross-pollinated plants derived and maintained by continuous close selfing in combination with the selection. Recurrent selection in the base populations would improve combining ability and performance or both. In any case, the frequencies of desirable alleles and allele combinations, i.e. genotypes, would be increased in the selected populations. Further, the magnitudes of these increases will become larger with the increasing number of selection cycles. Therefore, the chances of isolating superior inbreds will be much higher in the case of selected base populations than those of original base populations.

4.3 The Main Stages of Recurrent Selection

Recurrent selection is categorized into two main forms, namely inter-population and intra-population recurrent selections. The aim of inter-population in recurrent selection is to ensure that the performance of the two populations crossed is improved. When there is high level of heterosis in the main characters of a crop, it should, therefore, be subjugated in crosses. Moreover, first selfed (S1) offsprings for recombination units and full-sib or half-sib offsprings may be used by means of selection units in inter-population. On the other hands, intra-population in recurrent selection is used as an alternative where heterosis may be impossible. As such, full-sib or half-sib may be used for selection and recombination parts. The selection of whichever, first selfed (S1) or second selfed (S2) progenies as units, should be completely evaded because their reactions to selection for further breeding may result to substandard outcome than even with those with non-selfed descendants (Souza 2001). This type of breeding method comprises four different stages as presented in Fig. 4.1.

4.4 Types of Recurrent Selection

There are four types of recurrent selection schemes, differentiated based on how plants with the desired traits are identified which include (i) simple recurrent selection (SRS), (ii) recurrent selection for general combining ability (RSGCA), (iii) recurrent selection for specific combining ability (RSSCA) and (iv) reciprocal recurrent selection (RRS). Among these schemes, modified reciprocal recurrent selection is often applied in oil palm breeding programme (Rajanaidu et al. 2000, 2013; Soh et al. 2017).

4.4.1 Simple Recurrent Selection

This method involves selection of desirable genotypes and self-pollination. The process is sometimes referred to as phenotypic recurrent selection because the desirable plants are selected based on morphological traits and simple test. Simple recurrent selection (SRS) method is an extension of the mass selection which does not involve a tester or combining ability (Bangarwa 2017). In simple terms, the scheme for mass selection is the simplest of the selection schemes applicable to both cross- and self-pollinated crop species. The population to be improved by mass selection is space-planted, whereby a large number of suitable plants with superior phenotype are allowed to open-pollinate. The derived seeds from the selected plants are then harvested in bulk which completes the first cycle of selection. The bulk seed represents the improved population. This cycle is then repeated one or more times.

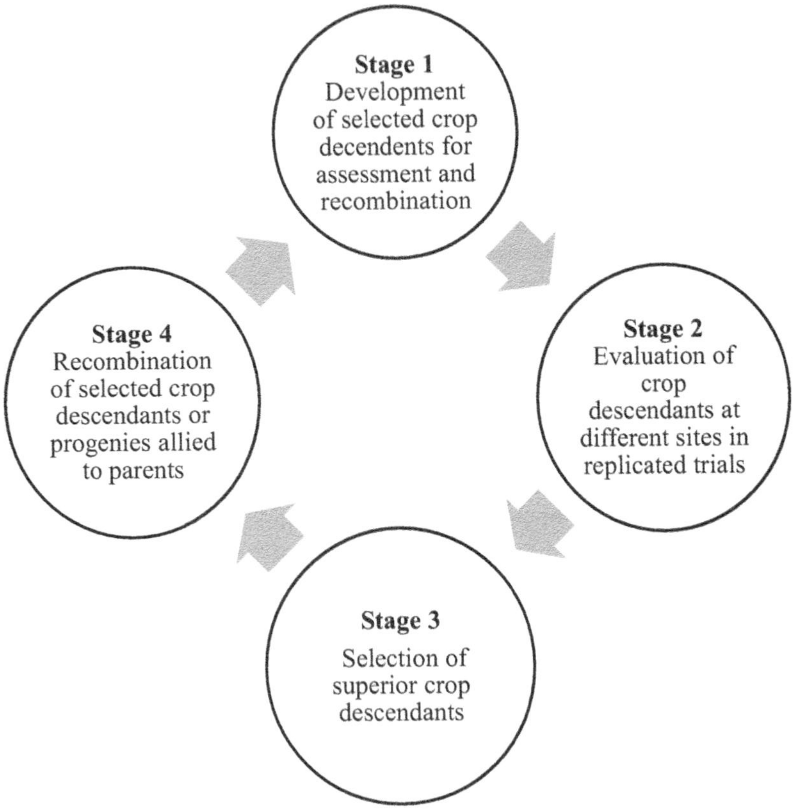

Fig. 4.1 Stages of recurrent selection

Mass selection is simple and easily implemented, but the open pollination of selected plants reduces its effectiveness. Further, inbreeding depression might set in which reduces the gains from selection if a significantly large number of plants is not selected. Locals in the African oil palm belt have initiated mass selection for desired traits including palm oil yield. Unfortunately, the yield increment from mass selection was very low (Bakoumé et al. 2016). Hence, SRS is useful only if the traits are highly heritable. SRS is based on two major groups. The group to be discarded and the group that is selected for further evaluation based on phenotypic score derived from selfed progeny plant (Acquaah 2009). In SRS, superior plants are selected from the original population. The selected plants are allowed to self-pollinate, and their seeds are harvested separately. The harvested seeds are then evaluated for desirable traits, and superior seeds are selected, while the rest are discarded. If the desirable traits are highly heritable, the individual plant progenies from the selfed seeds are planted in a crossing block during the following cycle of planting. In this case, all progenies are mated in all possible crosses, and equal amounts of seeds from all the crosses are mixed to form the selected version of the original population. This completes the first selection cycle; thus, one selection cycle is completed in two planting cycles. When this population is subjected to a second selection cycle, the scheme becomes a recurrent selection scheme. Therefore, a population is subjected to two or more selection cycles in a recurrent selection scheme.

The SRS scheme is useful for the improvement of the character having high heritability. The mean of the selected population for the trait under selection shifts in the direction of selection. Generally, there is a little decline in the genetic variability present in the selected population. The level of inbreeding remains low if there is sufficiently larger number of plants in each selection cycle.

4.4.2 Recurrent Selection for General Combining Ability (RSGCA)

In recurrent selection for general combining ability (RSGCA), selection is based on test-cross performance for improvement of general combining ability for a particular character of a population. This method is more effective for incomplete dominance character. Furthermore, the character in this scheme is mostly governed by additive gene action. Generally, an open-pollinated variety is used as a tester and requires three seasons or years for completion of each cycle of selection (Bangarwa 2017).

Jenkins (1940) proposed a scheme for the development of synthetic varieties from short-term inbreds. A short-term inbred is an inbred developed by a few generations of inbreeding. GCA is the ability of a line or plant to produce hybrids with superior average characteristics when crossed with several other varieties, plants, lines or strains or with a line that has broad genetic base. The GCA is believed to be a reflection of the additive gene action. Therefore, GCA is considered to be more relevant in the development of synthetic varieties.

The RSGCA scheme can be applied to the segregating generations of a hybrid or composite population, open-pollinated and synthetic varieties. In this scheme, the seeds used for replicated trial were produced by crossing the selected plants to a broad genetic base tester; the seeds produced is known as test-cross seeds. A tester is a line that is crossed with several lines or plants with the objective of estimating their GCA or SCA. The population to be subjected to selection is space-planted, and several superior plants are selected from this population. Each superior plant is self-pollinated as well as crossed with several random plants of the broad-based tester. The test-cross seeds from the entire selected plants are harvested separately and used for a replicated yield trial in the second year. The data from this trial are used to identify the test-cross progeny with superior performance. The selfed seeds from plants that produced superior test-cross progenies identified in the second year are planted as individual plant progenies in a crossing block. All possible crosses are made among these individual plant progenies, and equal amounts of seeds from all the crosses are mixed together to constitute the selected version of the population. This completes the first selection cycle. The composited seed is space-planted in the fourth year to initiate the first recurrent selection cycle. In the fourth year, the operations of the first year are repeated. Accordingly, in the fifth and sixth years, the operations of the second and third years, respectively, are also repeated. This completes the first recurrent selection cycles. The selection may be continued in more cycles. The use of RSGCA scheme improves the GCA of the population. In addition, it also improves the yielding ability of the selected population, at ~10% improvement per selection cycle (Corley and Tinker 2008).

4.4.3 Recurrent Selection for Specific Combining Ability

The recurrent selection for specific combining ability (RSSCA) was proposed by Hull in 1945 with a view to develop inbreds having superior combining ability with another specific inbred. This objective is of great relevance in the development of hybrid varieties since a large part of heterosis is believed to be due to SCA (Singh 2012). The procedure of RSSCA is the same as that of RSGCA except that RSSCA uses a narrow genetic base tester instead of a broad genetic base tester. The tester used in the case of RSSCA is an outstanding inbred valued as an excellent parent for the development of hybrid varieties. The RSSCA improves the population for combining ability with this inbred. Therefore, the inbreds isolated from the selected population are expected to combine well with this particular line, to produce hybrids with excellent performance (Acquaah 2009).

The population to be subjected to selection is space-planted, and several superior plants are selected. The selected plants are self-pollinated and simultaneously test-crossed with the inbred tester. The test-cross seeds are used for a replicated trial in the second year. This allows identification of test-cross progenies with superior performance. The selfed seeds from the plants producing the superior test-cross progenies are planted in a crossing block, and all possible crosses among the progenies are made. This completes the first selection cycle. These composite seeds represent the improved version of the original population. The seeds are space-planted in the fourth year, and the operations of the first year are repeated. Similar to RSGCA, the selection can be continued in more cycles.

One selection cycle of RSSCA is completed in three years. This scheme improves the population for SCA with the inbred used as the tester. It also improves the population for GCA, and it is estimated that 5% yield gain is improved in each selection cycle. The improved population is generally used for isolation of inbreds. These inbreds are expected to combine well with the inbred used as the tester. Further, these inbreds are used for hybrid variety development. Usually, the tester inbred will be used as one of the parents of the hybrid variety.

4.4.4 Reciprocal Recurrent Selection (RRS)

The RRS scheme was proposed by Comstock and his co-workers in 1949, with the objective of improving two different populations, for example, populations A and B, for combining well with each other. The two populations are subjected to selection at the same time. Further, one population, e.g. population A, serves as the tester for the plants selected from population B, and vice versa. As a result, the two populations are improved for combining ability with each other. In the first year, populations A and B are planted in separate fields. Several phenotypically superior plants are selected from population A, and they are self-pollinated. Each of these selected plants is also crossed with several random plants from population B; this constitutes test-cross. The selfed seeds and the test-cross seeds from all the selected plants are harvested separately. In the second year, the test-cross seeds are planted in a replicated trial to assess the performance of the test-cross progenies. The trial data allow the identification of superior test-cross progenies. In the third year, the selfed seeds from the plants producing these superior progenies are planted in individual plant progeny rows in a crossing block, all possible crosses are made among these progenies, and equal amounts of seeds from all the crosses are mixed (bulked). This completes the first selection cycle, and the bulked seeds represent the improved version of the original population A. These bulked seeds are planted in the fourth year, and the operations of the first year are repeated. This completes the first recurrent selection cycle. The selection can be continued for more cycles. Population B is subjected to the same procedure as described for population A. The only difference is each plant selected from the population B is crossed with several random plants from the population A, and the test-cross progenies produced in this way are used for replicated trial in the second year. The rest of the procedure remains the same as population A.

In general, RRS selects mainly for additive and dominance gene effects. The RRS scheme improves the yields of the two populations by approximately 4% per selection cycle. The GCA of the two populations is also improved. Further, the yield of the crosses between the populations increases by about 5% for every cycle of selection. The main reason for the increase is the improvement in the GCA of the two populations. The improved populations may be crossed to produce a hybrid that can be used as a synthetic variety. Alternatively, inbred lines may be isolated from the two improved populations, and they can be crossed to develop hybrid varieties.

4.5 Application of Recurrent Selection in Oil Palm Breeding

Systematic breeding for most annual and perennial hybrid crops has gravitated to two general schemes that are widely used by researchers, namely the modified recurrent selection (MRS) scheme (Rosenquist 1990; Soh 1999) and modified reciprocal recurrent selection (MRRS) scheme (Soh et al. 2017). The MRS focuses on establishing crosses with high GCA, whereas the MRRS enables the selection for SCA (Rajanaidu et al. 2013). GCA is defined as the average performance of a line in hybrid combination, whereas SCA refers to certain combinations in the population with better or lower performance than the average (Sritharan et al. 2017).

4.5.1 Modified Recurrent Selection Scheme (MRS)

The concept of MRS scheme was widely promoted by the breeding and genetic centre of Oil Palm Laboratory, a consortium that comprised of several oil palm private companies, namely Harisson and Crossfield (H&C), Guthrie, Pamol and Dunlop (Soh et al. 2017). This scheme has also been adopted and practiced by most breeding programmes across the major oil palm cultivating countries like Zaire (UniPamol), Thailand (Univanich Palm Oil), Colombia (Cenipalma), Malaysia (Sime Darby Plantation, KULIM Plantations Berhad, IOI Corporation Berhad, Malaysian Palm Oil Board (MPOB), Indonesia (SumBio, SMART) and Papua New Guinea (Dami Oil Palm Research Station) (Soh et al. 2017). The MRS scheme was adopted due to the initial need to produce a large amount of seeds to meet the rapidly expanding planta-tion demand and the scarcity of the available "West African *pisifera* (WAP)" pollen, which had to be developed through introgression of WAP/*tenera* (*T*) into the Deli to give rise to the Ulu Remis *teneras*/*pisiferas* (*P*). This went against the principles of clear separation of the heterotic parental populations to maximize hybrid vigour or "inter-origin effect" as adopted by Institut de Recherche pour Les Huiles et Oleagineux (IRHO) MRRS scheme (Soh et al. 2017).

This scheme involved the use of Deli *dura* (*D*) as a maternal parent in the commercial *tenera* hybrid seed production. The *pisiferas*, being female sterile (either bred or introduced), are primarily selected in line with their *tenera* sib performance in the $T \times T/P$ cross. The selected *pisifera* is progeny tested with selected *dura* mother palms. The commercial *tenera* hybrid seed production involves crossing between selected *dura* and *pisifera*. Rosenquist (1990) described MRS scheme as individual and family palm selection scheme because selection of parents is often done at family and individual levels. Since parental palms are selected at each breeding cycle, Soh et al. (2003) defined this as a form of recurrent selection. This MRS scheme exploits and emphasizes the effects of the GCA. Hence, more parental palms are usually selected from the MRS scheme as compared to the RSS scheme.

The main hypothesis of this scheme is the expression of the GCA within the population crosses (as expressed by the combining ability analysis or average means of the family or individual) that will be replicated within the inter-population hybrid crosses (Soh et al. 2003). The undue lengthy reliance on the Deli palm as the sole source of the *dura* maternal parent highlights one of the major limitations of the MRS scheme (Soh et al. 2017). Van der Vossen (1974) questioned the efficiency of the scheme because of the exclusive utilization of the Deli as an external pam. As a result, Hartley (1988) observed that the exclusive use of the Deli population as the female parent in the seed production and maintaining it as pure lines might unnecessarily cause dependence on a limited inbred population. Nonetheless, with few exceptions, Deli *dura* remains to be utilized 25 years later as the sole female parent for seed production in all parts of the world where oil palm is cultivated (Corley and Tinker 2003). A simplified sketch of the reciprocal recurrent selection programme in operation at Centre de Coopération Internationale en Recherche Agronomique pour le Développement (CIRAD) is as shown in Fig. 4.2.

4.5.2 Modified Reciprocal Recurrent Selection Scheme (MRRS)

The MRRS scheme has been adopted by the breeding programmes in some West African

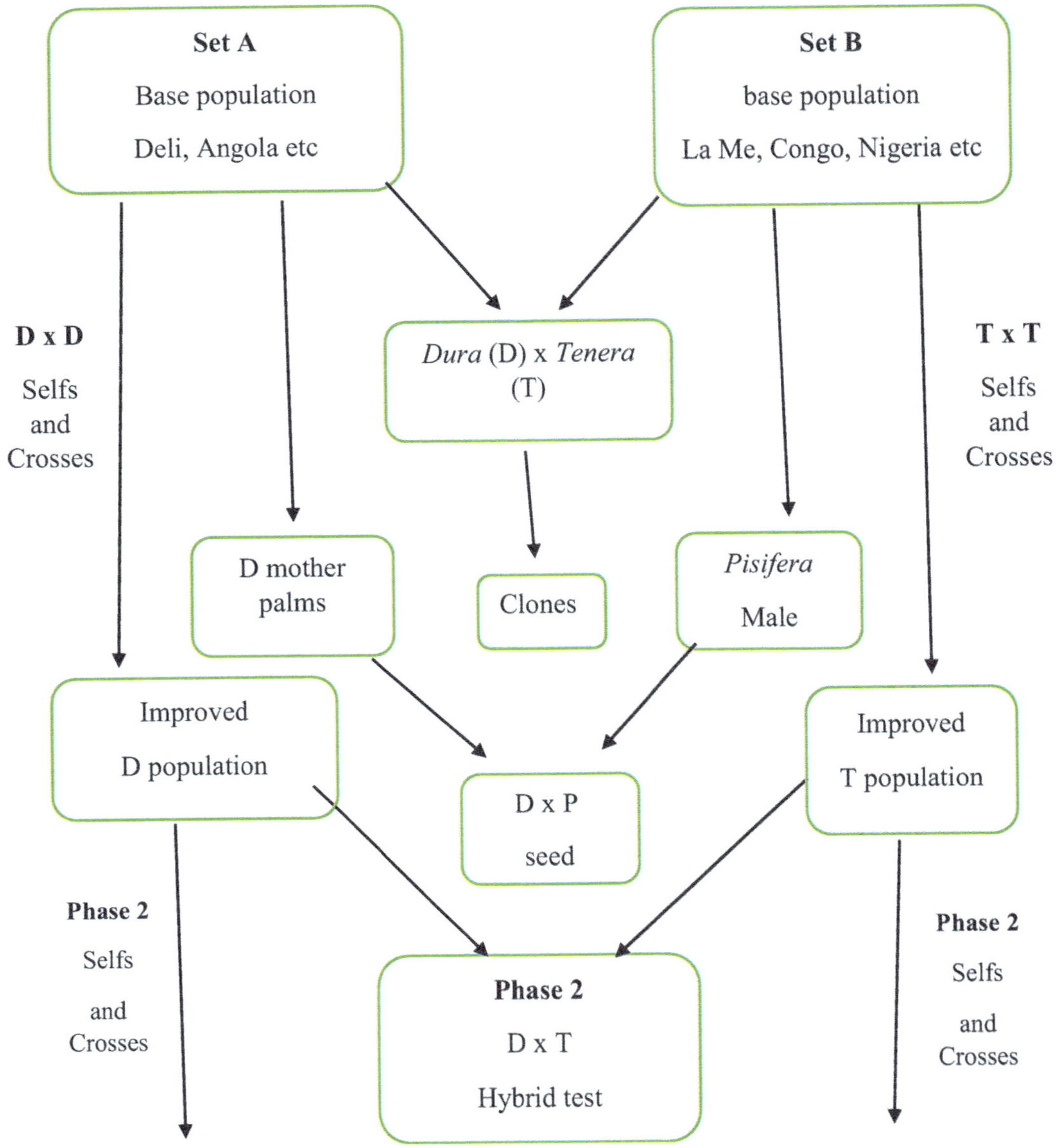

Fig. 4.2 Simplified sketch of CIRAD recurrent reciprocal selection programme (Modified from Corley and Tinker 2016)

research institutes, namely Nigerian Institute for Oil Palm Research (NIFOR), Ghana Oil Plan Research Institute (GOPRI) and Centre de Coopération Internationale en Recherche Agronomique pour le Développement (CIRAD) coordinated programmes established in Cote d'Ivoire, Cameroon, Benin, Niger and Guinea-Bissau. It has also been adopted in South East Asia by Indonesian Oil Palm Research Institute (IOPRI) and SOCFINDO in Indonesia based on breeding populations derived from CIRAD such as Deli, Angola, Yangambi, La Mé and Yocobue. Other Indonesian oil palm agencies, for example, SampoernaAgro, SAIN, Tania Selatan and Bakrie also adopted this approach strategically, to kick start their seed production and breeding programmes utilizing breeding populations derived from Agriculture Services and

Development (ASD), Costa Rica, via seed purchase or franchising. MRRS has also crept into breeding programmes in Malaysia, particularly in Applied Agriculture Resources Sdn Bhd (AAR) and MPOB. The first RRS in oil palm breeding programme was implemented at Yangambi by Institut National pour l'Étude Agronomique du Congo Belge (INEAC) between 1959 and 1966 in La Mé under the name "Block 500". Parents derived from both the International Experiment and new introductions were crossed, resulting in 529 progenies (Soh et al. 2017). The initial stage of the RSS programme consists of a diallel cross among six *teneras* (Pichel 1956).

North Carolina Mating (NCM) 1 breeding design can estimate the specific and general combining abilities of the *pisiferas* (Rafii et al. 2002). The GCA can be assessed by looking at the mean performance of a *pisifera* to several Deli *duras*. If the mean is high, the GCA of the *pisifera* is considered good and vice versa. Several major oil palm breeding agencies around the world implement RRS for keeping the *pisifera* and *dura* populations distinct. Sib-crossing and selfing are usually included in this scheme to obtain parents for seed production or for the next generation of crosses (Corley and Tinker 2003). RRS was developed from maize breeding programme (Okporie et al. 2013; Comstock et al. 1949) to exploits both the GCA and the SCA effects. The adaptation of such scheme in oil palm breeding presumably come about from the fact that there is greater possibility for a heterotic yield from crosses between Deli *dura* with African *tenera* or *pisifera* (Soh et al. 2017).

This attractive MRRS scheme has three main advantages. Firstly, selection of selfed parents for commercial hybrid seed production and for a more advanced breeding programmes is both derived from the progeny tests (*tenera* and *dura* grandparents) of the potential commercial inter-population hybrids. Secondly, the RRS scheme is made up of two distinct phases, viz. the "recombinant stage" that involves outcross which runs parallel to each other. This stage allows the accumulation of desirable alleles (both additive and non-additive). It also ensures the maintenance of genetic variability for sustainable long-term improvement. The "within hybrid improvement stage" facilitates short-term commercial exploitation of the most suitable test-crosses. This enhances recurrent selection involving parental selfs or sibs. Thirdly, once there are available data on the inter-population hybrid, production of the commercial hybrid seeds can be carried out using the *duras* and *pisiferas* that are generated from the selfs or sibs of the respective parents, which are eventually planted simultaneously as the progeny tests (Jacquemard et al. 1981). The major drawback of this scheme is the requirement of a large area for field planting. To generate 3–4 million improved hybrid seeds from the reproduction of selected top 15% of the crosses, a total farm area of 600 ha is needed to lay down 180 parental selfs and about 500 crosses within a period of cultivation ranging from 15 to 25 years (Soh 1999).

The major setback of RRS scheme is the prolonged "generation time" since inter-population crosses essentially need to be tested before the next cycle parents are selected. Two generations can effectively be occupied in one cycle. If the selfing of all parents is produced simultaneously with the inter-population test-crosses, then the prolonged generation time can be drastically reduced. However, the reduced time is at the expense of a greater land area for field trials. In the CIRAD breeding programme, selected parents are selfed simultaneously, while the test-crosses are carried out few years later (Durand-Gasselin et al. 2000). Figure 4.3 shows the modified reciprocal recurrent selection in oil palm breeding.

4.6 Progression of Oil Palm Through Recurrent Selection

The first application of RRS scheme in oil palm programme was implemented by INEAC at Yangambi in the Democratic Republic of Congo. Diallel cross among six *teneras* constituted the initial phase of the programme (Pichel 1956). The crossing was aimed at combining high-yielding T×T from the RRS scheme which could be reproduced through the combination of *duras* from

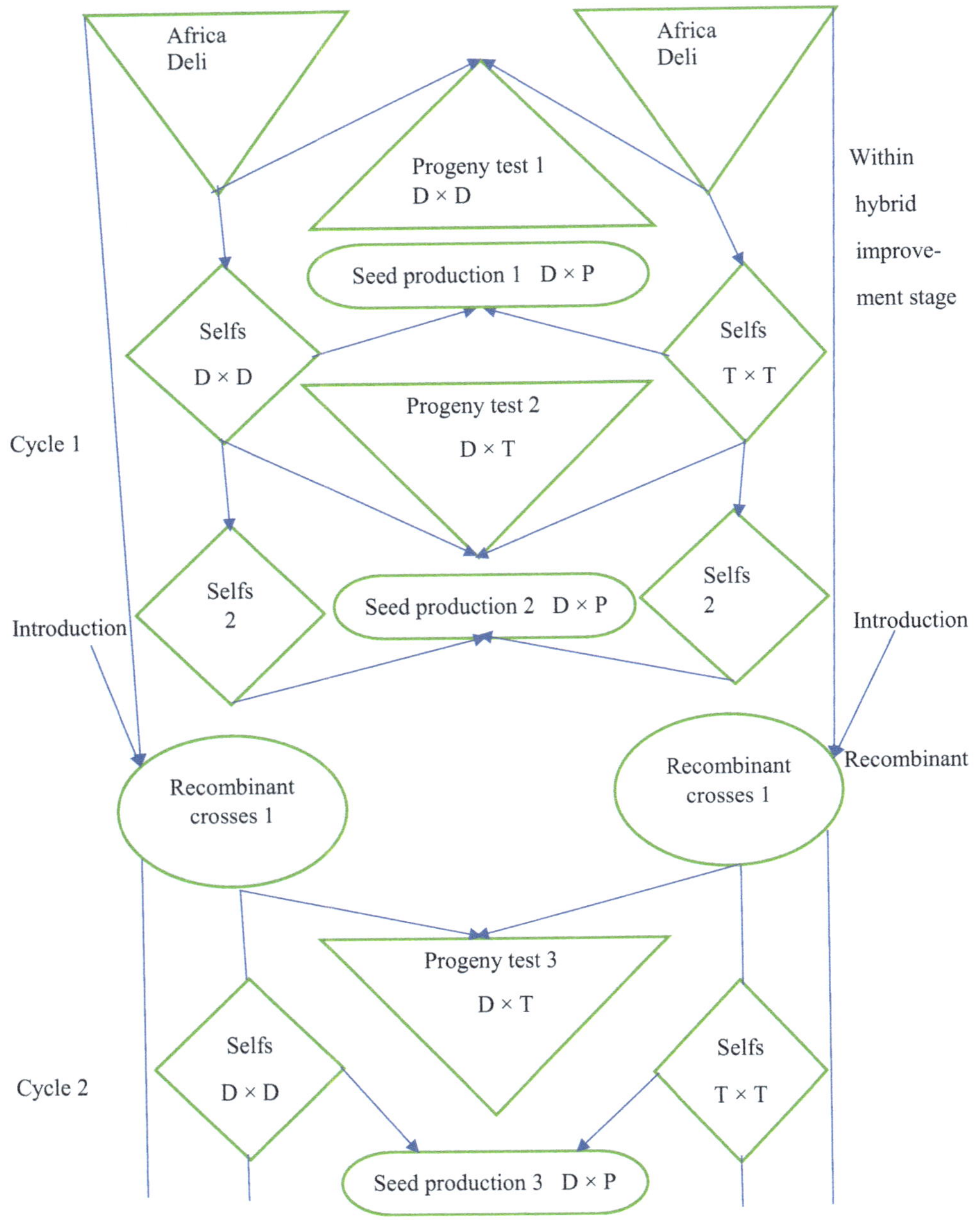

Fig. 4.3 Modified reciprocal recurrent selection in oil palm breeding (Modified from Rajanaidu et al. 2000)

one of the selfed *tenera* progenies with other selfed *pisifera* progenies (Gascon and de Berchoux 1964). Subsequently, both Nigerian Institute for Oil Palm Research (NIFOR) in Nigeria and Institut de Recherche pour les Huiles et Oléagineux (IRHO) in Ivory Coast apply this RSS scheme in their oil palm breeding programmes (Sparnaaij et al. 1963; Gascon and de Berchoux 1964).

The first selection phase of the reciprocal recurrent selection led to yield increase of 15% (Gascon et al. 1981) followed by additional improvement of 15% during the second cycle population (Cochard et al. 1993; Gascon et al. 1988). Furthermore, with acceptable growing conditions in Malaysia, there was an increase in oil yield of 1.3 t/ha in 1950 to 5.4 t/ha in 1990.

Within these periods, nearly half of the improvement was contributed by the improved parental lines (Davidson 1991). Crude palm oil, which accounts for about 47% market portion, has gained a lot of improvements through germplasm development via recurrent selection (Barcelos et al. 2015).

In Indonesia and Ivory Coast, the populations derived from the second cycle of RRS were laid down between 1975 and 2001. The results showed meaningful impact on the commercial seed production (Ngando-Ebongue et al. 2012). Durand-Gasselin et al. (2000) reported that in Ivory Coast, with inappropriate conditions, the yield potential of the oil palm materials planted was 2.9 t/ha/yr in 1950 and increased to 4.1 t/ha/yr in 1998. The Africa Industry News (2016) projected a total oil palm capacity of 600,000 t/yr in 2020 for Ivory Coast, as a result from the effective recurrent selection imposed in a 19-year period of breeding cycle. In 2005, a collaboration between Socfindo, Murrin and Siat groups saw the setting up of the third cycle of field trials in Ecuador, Indonesia and Nigeria.

PalmElit (previously known as IRHO/CIRAD) RRS programme started with a very narrow genetic base, but continuous work towards introducing new genetic materials into both RRS groups (SCA and GCA) was undertaken (Meunier and Gascon 1972; Gascon et al. 1988). Several populations resulting from prospections (Yocoboué, Nigeria, Widikum) and exchange programmes (Angola, Lobé, Aba, Calabar, Sibiti, Yangambi, and various Deli BPROs) were introduced in La Mé and Pobé between the 1960s and 1980s. These populations were assessed for both their individual trait data and breeding values in order to select and introduce the best parents into the breeding programme. Two cycles of introgression, testing and recombination were carried out involving Angola populations (Adon et al. 1998). In a joint breeding programme, PalmElit–Socfindo has established a new genetic block, to test 149 Group A and 123 Group B parents. This block is exceptionally unique because of the introduction of some new progenitors that have never been used before in any joint breeding programme by PalmElit–Socfindo (Soh et al. 2017).

In Benin, planting of the third-cycle trials began in 2010 and ended in 2016. These trials focus on testing selfs and recombinants of the best parents from the second cycle that exhibit tolerance or good resistance to Fusarium wilt. In Nigeria and Ecuador, testing of 255 Group A parents and 145 Group B parents is being conducted on 800 ha (Soh et al. 2017).

4.7 Conclusion

The oil palm which predominantly ascends as a perennial crop with genetic improvement through recurrent selection by oil palm breeders is projected to last for about 25–30 years of substantial yield output. The oil palm productivity can exceed the expected projection if super high-yielding planting materials are adopted. It is therefore essential for oil palm breeders to assure performance of quality planting materials to satisfy the increasing demand for vegetable fats and oils. The application of modified recurrent selection scheme in oil palm would allow introduction of new breeding populations towards generating new planting materials with value-added products. This will also help increase palm oil value as well as maintain considerable genetic diversity for selection gains and sustainability in future.

References

Acquaah G (2009) Principles of plant genetics and breeding. Wiley, Hoboken

Adon B, Baudouin L, Durand-Gasselin T, Kouamé B (1998) Utilisation de matériel non amélioré pour la sélection du palmier à huile: l'origine Angola. Plantations, Recherche, Développement 5(3):201–207

Africa Industry News (2016) https://www.palmoilextractionmachine.com/news/industry_news/86.html

Bakoumé C, Ngando Ebongué G, Ajambang W, Ataga CD, Okoye MN, Enaberue LO, Konan JN, Allou D, Diabate S, Konan E, Etta CE (2016) Oil palm breeding and seed production in Africa. Proceedings of the international seminar on oil palm breeding and seed production and field visits, Kisaran, Indonesia, 29–30 September 2016

Bangarwa KS (2017) Recurrent selection-definition and types: agriculture. Biotech Articles. https://www.biotecharticles.com/Agriculture-Article/Recurrent-Selection-Definition-and-Types-4144.html. Accessed July 2018

Barcelos E, Rios SDA, Cunha RN, Lopes R, Motoike SY, Babiychuk E et al (2015) Oil palm natural diversity and the potential for yield improvement. Front Plant Sci 6:190

Cochard B, Noiret JM, Baudouin L, Flori A, Amblard PH (1993) Second cycle reciprocal recurrent selection in oil palm, *Elaeis guineensis* Jacq. Results of Deli × La Mé hybrid tests. Oléagineux (Paris) 48(11):441–451

Comstock RE, Robinson HF, Harvey PH (1949) A breeding procedure designed to make maximum use of both general and specific combining ability 1. Agron J 41(8):360–367

Corley RHV, Tinker B (2003) The oil palm, 4th edn. Blackwell Science LTD, Oxford, p 562

Corley RHV, Tinker PB (2008) The oil palm. Wiley, New York

Corley RHV, Tinker PB (2016) The oil palm, 5th edn. Wiley, New York

Davidson L (1991) Management for efficient cost-effective and productive oil palm plantations. In: PORIM international conference progress prospects challenges towards the 21st century, Kuala Lumpur (No. D-0565). Unilever Plantation, September 9–14

Durand-Gasselin T, Kouame Kouame R, Cochard B, Adon B, Amblard P (2000) Diffusion variétale du palmier à huile (*Elaeis guineensis* Jacq.). OCL. Oléagineux Corps gras Lipides 7(2):207–214

Gascon JP, De Berchoux C (1964) Caractéristiques de la production d'Elaeis guineensis (Jacq.) de diverses origines et leurs croisements. Application à la sélection du palmier à huile. Oléagineux 19(2):75–84

Gascon JP, Jacquemard JC, Houssou M, Boutin D, Chaillard H, Kanga Fondjo F (1981) La production de semences sélectionnées de palmier à huile. Oléagineux 36(10):475–486

Gascon JP, Guen VL, Nouy B, Kamga F (1988) Results of second cycle recurrent reciprocal selection trials on oil palm Elaeis guineensis Jacq. Oléagineux 43(1):1–7

Hallauer AR, Darrah LL (1985) Compendium of recurrent selection methods and their application. Crit Rev Plant Sci 3(1):1–33

Hartley CWS (1988) The oil palm, 3rd edn. Longman, London/New York

Hull FH (1945) Recurrent selection for specific combining ability in corn 1. Agron J 37(2):134–145

Jacquemard JC, Meunier J, Bonnot F (1981) Genetic study of the reproduction of an *Elaeis guineensis* oil palm cross. Oleagineux 36(7):343–352

Jenkins MT (1940) The segregation of genes affecting yield of grain in maize. J Am Soc Agron 32:55–63

Kushairi A, Mohd Din A, Rajanaidu N (2011) Oil palm breeding and seed production. In: Mohd Basri W, Choo YM, Chan KW (eds) Further advances in oil palm research. MPOB, Bangi, pp 47–101

Meunier J, Gascon JP (1972) Le schéma général d'amélioration du palmier à huile à l'IRHO. Oléagineux 27(1):1–12

Ngando-Ebongue GF, Ajambang WN, Koona P, Firman BL, Arondel V (2012) Oil palm. In: Technological innovations in major world oil crops, vol 1. Springer, New York, pp 165–200

Okporie EO, Chukwu SC, Onyishi GC (2013) Development of high protein populations of maize (Zea mays L.) from three cycles of reciprocal recurrent selection. IOSR J Agric Vet Sci 3(2):22–26

Pichel R (1956) L'amélioration du palmier à huile au Congo Belge. Bull Agron 14:59–66

Rafii MY, Rajanaidu N, Jalani BS, Kushairi A (2002) Performance and heritability estimations on oil palm progenies tested in different environments. J Oil Palm Res 14(1):15–24

Rajanaidu N, Kushairi A, Rafii M, Mohd Din A, Maizura I, Jalani BS (2000) Oil palm breeding and genetic resources. Advances in oil palm research, pp 171–237

Rajanaidu N, Ainul MM, Kushairi A, Mohd Din A (2013) Historical review of oil palm breeding for the past 50 years—Malaysian journey. In: Proceedings of the international seminar on oil palm breeding—yesterday, today and tomorrow, Kuala Lumpur, Malaysia, pp 11–28

Rosenquist EA (1990) An overview of breeding technology and selection in Elaeis guineensis. In: 1989 PORIM international palm oil development conference. Agriculture, Kuala Lumpur, Malaysia (No. L-0076). PORIM, 5–9 September 1989

Singh BD (2012) Plant breeding: principles and methods. Kalyani Publishers, New Delhi

Soh AC (1999) Breeding plans and selection methods in oil palm. In: Proceedings of symposium on the science of oil palm breeding. Montpellier, Francia (No. L-0435). PORIM

Soh AC, Wong G, Hor TY, Tan CC, Chew PS (2003) Oil palm genetic improvement. Plant Breed Rev 22:165–220

Soh AC, Mayes S, Roberts J (2017) Oil palm breeding: genetics and genomics. Oil palm breeding: genetics and genomics. https://doi.org/10.1201/9781315119724

Souza CL Jr (2001) Melhoramento de espéciesalógamas. In: Nass LL, Valois ACC, Melo IS, Valadares-Inglis MC (eds) Recursosgenéticos e melhoramento: plantas. Fundação MT, Rondonópolis, pp 159–199

Sparnaaij LD, Menendez T, Blaak G (1963) Breeding and inheritance in the oil palm (ElaeisguineensisJacq.) Part I the design of a breeding programme. J West Afr Inst Oil Palm Res 4:126–155

Sritharan K, Subramaniam M, Arulandoo X, Yusop MR (2017) Yield and bunch quality component comparison between two-way crosses and multi-way crosses of DxP oil palm progenies. Sains Malaysiana 46 (9):1587–1595

Van der Vossen HAM (1974) Towards more efficient selection for oil yield in the oil palm (*Elaeisguineensis* Jacquin). Thesis, Wageningen University

Oil Palm Tissue Culture: Fast Tracking Elite Commercial Lines

5

Zamzuri Ishak, Ahmad Tarmizi Hashim, Samsul Kamal Rosli, Dalilah Abu Bakar, Siew Eng Ooi, Naqiuddin Mohd, and Meilina Ong-Abdullah

Abstract

Enhancing oil palm productivity through land expansion is no longer a viable option, and thus, improving yield per hectare is the utmost priority for the Malaysian Oil Palm Industry. In order to produce improved and high-yielding planting materials, the Malaysian Palm Oil Board (MPOB) has leveraged its rich genetic resource established since 50 years ago through extensive bioprospection as well as augmenting its cloning facilities to fast-track the breeding programmes. Cloning via the tissue culture system is by far the most viable approach to vegetatively propagate elite palms. After more than 33 years of research and development, reliable protocols for both solid and liquid culture systems coupled with various innovative technologies were established. The SureSawit™ Karma biomarker, which is related to clonal abnormality, was developed as a tool for quality assurance of clonal production. To date, clonal trials have been extensively conducted, and the results have been encouraging. Clones such as CPS1, CPS2 and a few others recorded superior performance in comparison with D×P standard crosses. Tissue culture has also enabled selection of clones with special characteristics such as virescens, compactness, high carotene content and other traits for future recloning.

5.1 Introduction

Oil palm (*Elaeis guineensis* Jacq.) planting materials are being sexually produced by crossing *dura* and *pisifera* (D×P), resulting in the *tenera* offspring. While the D×P seed materials have contributed to quantum leaps in yield improvements, there still exists considerable variations for exploitations through vegetative propagation via tissue culture. This method enables true-to-type reproduction of the best genotypes which is very much desired by the plantation community.

Vegetative propagation of oil palm via tissue culture technique took place as early as 1960s through 1970s. The early success of ramets (tissue culture derived planting materials) production was seen in the 1970s (Jones 1974; Rabéchault and Martin 1976). This success inspired Malaysian Palm Oil Board, MPOB (then Palm Oil Research Institute Malaysia, PORIM), and many other oil palm organizations to proceed with large-scale production of ramets. In the mid-1980s, the first report on floral abnormality in clonal oil palm (Corley et al. 1986), known as mantling, had caused a commotion among oil

Z. Ishak · A. T. Hashim · S. K. Rosli · D. Abu Bakar · S. E. Ooi · N. Mohd · M. Ong-Abdullah (✉)
Malaysian Palm Oil Board, No. 6, Persiaran Institusi, Bandar Baru Bangi, 43000 Kajang, Selangor, Malaysia
e-mail: meilina@mpob.gov.my

© Springer Nature Switzerland AG 2020
M. Ithnin and A. Kushairi (eds.), *The Oil Palm Genome*, Compendium of Plant Genomes, https://doi.org/10.1007/978-3-030-22549-0_5

Fig. 5.1 Normal and mantled fruit form of the oil palm. Floral abnormality found in tissue culture derived planting materials will lead to the formation of mantled fruits which have significantly reduced oil content, or at worse prevents the fruit formation

palm tissue culturists (Fig. 5.1). Many laboratories reduced their production and produced only enough plantlets for field evaluation.

After a couple of decades, as more data and deeper understanding obtained on the tissue culture process and its associated problems, ramets' producers regained their confidence to bulk-produce clonal materials for field plantings. Despite recording significantly reduced abnormality occurrence, the large-scale production of ramets has been impeded as a repercussion on the modified approach applied to the formulation of tissue culture media and its protocols. The low rate of embryogenesis, ranging from an average of 3–6% (Rajanaidu et al. 1997; Corley and Tinker 2003), had been adding insult to injury. Of that, about 50% of the embryoids failed to establish (Wooi 1995). Against all odds, most laboratories have now established: (i) enhanced tissue culture media and protocols and (ii) field evaluation of their clonal materials. The floral abnormality observed in the clonal trials of the later clonal production gave an abnormality rate of less than 5% compared to the earlier production, whereby the rate exceeded 5% (Tan et al. 2003). In order to keep this rate low, reliable tissue culture procedures and stringent culling at various cloning stages (Maheran et al. 1995; Simon et al. 1998) as well as the use of a wider range of ortets (Tan et al. 2003) need to be practiced. To date, the abnormality rate still remains less than 5% as reported by tissue culture laboratories (Hashim et al. 2018). Performance wise, clonal plantlets derived from selected ortets have been reported with higher yield compared to commercial D×P seedlings (Khaw and Ng 1997). In general, clones yielded approximately 20% more than seedling standards (Rohani et al. 2000; Corley and Tinker 2003; Tan et al. 2003). However, there are still issues and challenges to be addressed as covered in this review.

5.2 Tissue Culture as a Fast Track Breeding Programme

According to a survey conducted by MPOB, yields from areas planted with oil palm in Malaysia have been stagnating, hovering at 18–20 tonnes Fresh Fruit Bunches (FFB) per hectare per year (t/ha/yr) and Oil Extraction Rate (OER) of 19–20% since 1994 with the national average oil yield (OY) of less than 3.7 t/ha/yr (Basiron 2007). Furthermore, in recent years since 2013–2016, the Oil Yield (OY) has further dropped from 3.7 to 3.51 t/ha/yr (MPOB Pocketbook 2017). The stagnating and declining yield is not new as the issue has been brought up in several forums for many years.

To address this issue, the oil palm breeders continue to establish improved planting materials through introgression of existing and germplasm materials (Rajanaidu and Jalani 1999). Using this technique, high-yielding materials have been propagated; however, despite rigorous selection, the problem of segregation in the seed-derived progenies still persists. Another bottleneck of this conventional method of oil palm breeding is the fact that it is tedious and time consuming. In addition, expansion of land as a means of increasing productivity is not feasible and sustainable, and hence, increasing yield and profitability on the existing plantation area is the only viable alternative. According to Corley and

Tinker (2003), the genetic potential of palm oil yield can reach up to 18.5 t/ha/yr, and at the moment, this figure is far from achievable with the national average never recorded above 4 t/ha/yr OY per year. Therefore, to fast-track the progress and to capture the maximum potential of a selected genotype, vegetative propagation through tissue culture is the best option available to the industry.

Tissue culture or micropropagation is a technique used to propagate and maintain plant cells, tissues or organs on a culture medium of known and controllable nutrient composition under sterile conditions. Plant tissue culture is widely used to produce clones of economically important crops, including the oil palm. In oil palm breeding programmes, tissue culture is incorporated to shorten the time taken to develop commercial planting materials. Via cloning, it is expected that multiplication of individuals with desired economic traits can be expedited. The process of oil palm tissue culture is illustrated in Fig. 5.2. Tissue culture offers an extremely valuable strategy which is not only limited for multiplying high-yielding individuals but also to produce elite clones with economically and agronomically desirable traits such as high bunch index, low height increment, more compact palm, disease resistance, among others. However, the main focus in this review is on cloning of high-yielding planting materials.

5.3 Performance of High-Yielding Clonal Palms

Clonal palms have been field tested by MPOB and few other agencies in various plantation areas. As mentioned previously, clones generally show superior performance compared to commercial D×P planting materials. For instance, the cumulative FFB of 6–7-year-old clones planted in Sarawak was reported 10–37% higher compared to the standard D×P (Simon and Koh 2005). Similar trend was observed in a plantation in Sabah, Malaysia, the cumulative FFB for 9–10-year-old clones exceeded those of D×P by 28–55%. Agencies such as FELDA (Roowi 2010) and UPB (Sharma 2006) also reported increased in FFB by 20% or more. In general, clones produced on average 48 t/ha of FFB, which is 44% increase over D×P (Simon and Koh 2005).

5.3.1 MPOB Clonal Performance

Performance of MPOB clones in few field trials from different locations in Malaysia is as follows.

5.3.1.1 Field Trials of MPOB Clones at MPOB Station

Table 5.1 shows seven years data of mean FFB yield per hectare from third to ninth year after planting at MPOB Station in Keratong. In this respect, all clones were superior to the D×P control ranging from 3 to 38%. Clone P90 outperformed the D×P by 38% while both P56 and P143 by 32%. The average FFB increment of the ten clones over D×P was 21%. In comparing individual clones at the ninth year, P75, P90 and P56 produced 34.8, 34.3 and 31.2 t/ha FFB, respectively. Based on the FFB means and the % oil to bunch (O/B) obtained, OY for all clones were higher than the D×P from 7 to 48% with an average of 20% (Zamzuri 2004).

In terms of FFB yield per hectare, both clones and D×P started to increase their yields at the fifth year after planting. The clones exceeded D×P steadily throughout the years except at the seventh year (in 2001) where the D×P showed a dramatic increase in yield to more than 30 t/ha/yr but then dropped to 20 t/ha/yr in the following two years. However, the clones remained stable between 20 and 25 t/ha/yr.

5.3.1.2 Field Trials of MPOB Clones at Private Estates

(a) *JC Chang Plantations in Sabah*

This trial was planted on March 1997 near Lahad Datu in Sabah. Two clones (P149 and P164) were compared with D×P commercial palms. The total number of clonal materials was 720 palms. Only 0.3% palms were found mantled initially but later mostly reverted to normal fruiting. Table 5.2 shows the performance of

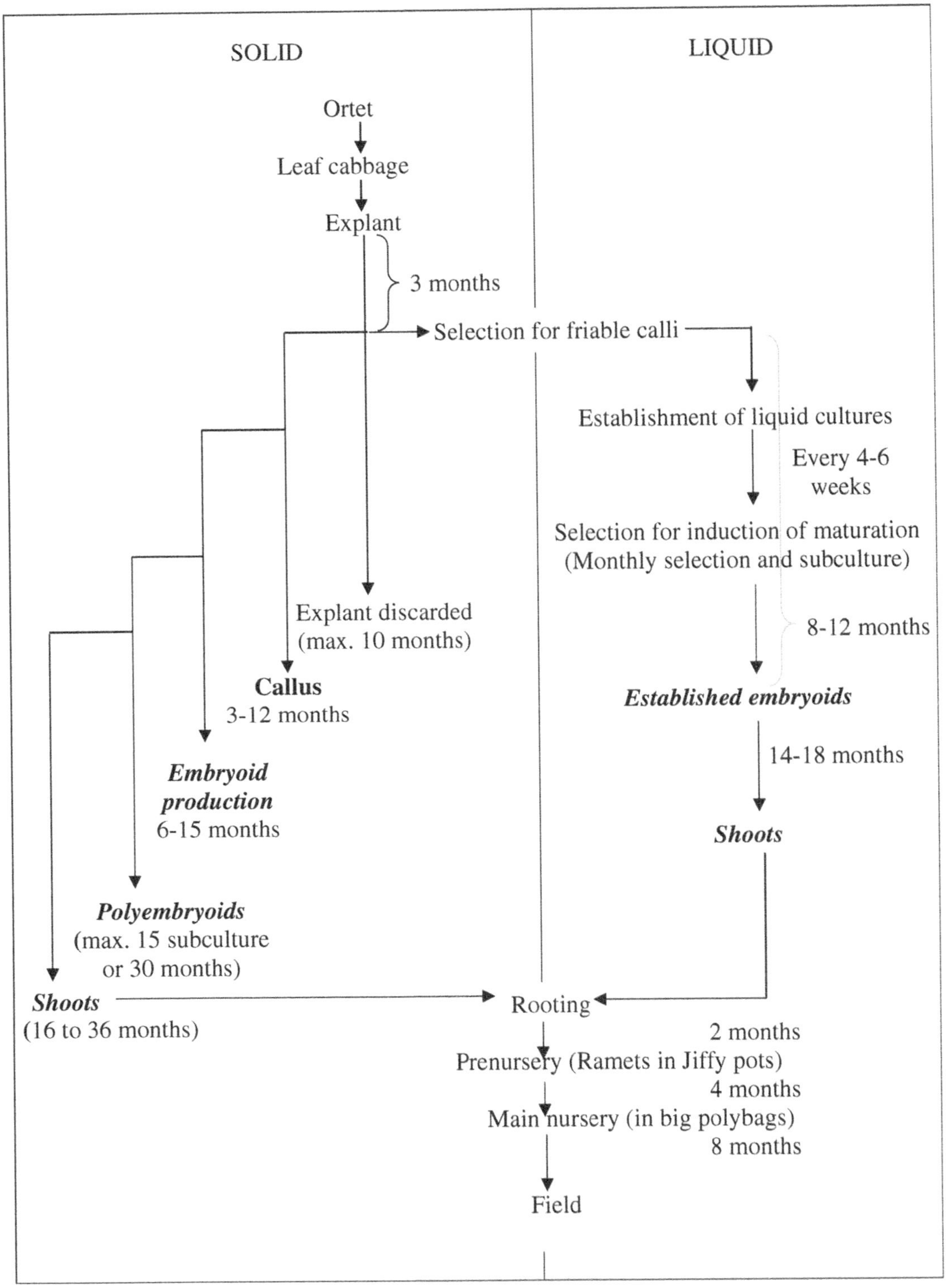

Fig. 5.2 MPOB oil palm tissue culture protocol with approximate time scale. The figure is adapted from Tarmizi et al. (2018)

Table 5.1 FFB, O/B and OY from third to ninth year after field planting

Clone no.	FFB (t/ha/yr)							Mean FFB	Incr[b] (%)	O/B (%)	OY t/ha	Incr[a] (%)
	Years after planting											
	3	4	5	6	7	8	9					
P56	4.88	10.41	11.51	15.16	28.43	25.74	31.19	18.19	32	24.07	4.38	26
P57	7.31	7.08	11.67	14.27	26.37	20.91	23.85	15.92	16	23.84	3.80	9
P75	6.90	8.01	16.67	8.80	14.84	15.16	34.79	15.02	9	25.61	3.85	11
P79	8.40	6.47	17.60	14.06	26.03	20.84	18.71	16.01	17	26.39	4.23	21
P90	8.72	9.49	15.33	16.80	24.29	23.32	34.26	18.89	38	27.28	5.15	48
P123	7.69	6.31	20.15	11.53	20.39	15.13	22.18	14.77	8	27.15	4.01	15
P126	6.55	9.65	22.35	21.22	22.95	24.65	18.35	17.96	31	23.91	4.29	23
P127	9.50	7.42	6.04	19.00	38.99	28.56	14.30	17.69	29	21.04	3.72	7
P135	6.43	9.27	13.26	17.52	21.82	15.63	14.82	14.11	3	26.97	3.80	9
P143	8.29	11.84	22.59	17.06	24.36	20.84	22.06	18.15	32	24.71	4.48	9
Clone mean	7.47	8.60	15.72	15.54	24.85	21.08	23.45	16.67	21	25.10	4.17	18
SC	5.89	5.13	9.16	12.72	32.61	17.44	13.19	13.73	–	25.35	3.48	–
Incr (%)[c]	27	68	72	22	−24	21	78	21				

SC Standard cross D×P
[a]OY increment of clones against SC, [b]FFB increment of clones against SC, [c]FFB increment of multiple clones against SC

Table 5.2 FFB yields (t/ha/yr) of clones against SC from JC Chang Plantations, Sabah, Malaysia

	Age of clones/palms (years)										
	5	6	7	8	9	10	11	12	13	14	15
MPOB Clones	32.01	27.35	31.06	32.46	30.00	36.26	34.55	32.60	32.68	37.03	29.65
SC[a]	25.84	24.22	26.86	26.17	26.54	30.93	29.23	26.09	30.93	30.37	23.93
Incr (%)[b]	23.88	12.92	15.64	24.04	13.04	17.23	18.20	24.95	24.95	21.93	23.90

[a]Standard cross D×P, [b]FFB increment of clones against SC

MPOB clones with an average FFB increment of 20.06%. The oil to bunch (O/B) ratio of the clonal materials in this plantation was 28.5% as compared to 21.2% of standard D×P materials (Details of O/B and OY data were not provided by the estate)

(b) *Serijaya Industri Sdn Bhd in Sabah*

TSH has planted over 900 hectares of MPOB clone P456 and the standard cross D×P at Gunung Rara Estate, Kalabakan, Sabah. In this trial, clone P456 showed an outstanding FFB yield performance over D×P material even at early age of production and at high altitude area of Gunung Rara estate in Kalabakan, Sabah as summarized in Table 5.3. O/B and OY data were not provided by the estate.

(c) *United Plantations Berhad, Bagan Datuk, Perak*

United Plantations Berhad has conducted a trial with D×P standard planting materials obtained from various plantation agencies to be compared with two MPOB clones. A trial plot was planted

Table 5.3 FFB Yield (t/ha/yr) of P456 against D×P standard cross at Gunung Rara Estate, Serijaya Industries Sdn Bhd, Sabah, Malaysia

	Age of clones/palms (years)		
	4	5	6
Clone P456	24.53	29.75	30.21
SC[a]	15.86	20.43	21.05
Incr (%)[b]	54.7	45.6	43.5

[a]Standard cross D×P, [b]FFB increment of clones against SC

Table 5.4 Performance of MPOB clones against agencies' D×P at the fifth year after planting in Bagan Datuk, Perak, Malaysia

Agency	FFB (t/ha/yr)	O/B (%)	OY (t/h/yr)
MPOB P126	24.23	27.09	6.56
MPOB P456	33.03	31.80	10.50
Agency A	27.44	29.41	8.07
Agency B	28.43	27.19	7.73
Agency C	30.86	28.34	8.74
Agency D	29.57	28.54	8.44
Agency E	29.20	24.73	7.22
Agency F	29.11	22.97	6.69
Agency G	26.25	25.47	6.69
Agency H	30.42	26.47	8.05
Agency I	28.29	26.69	7.55
Agency J	29.41	23.70	6.97

with D×P progenies from ten agencies (Agency A-J) and two MPOB clones in 1999. Table 5.4 shows the yield record at fifth year after planting. It is obvious from the table that MPOB clone P456 outperformed D×P of all agencies in almost every parameters measured. The second MPOB clone, P126, although seemed not to be performing well against D×P, has a potential to be used in high-density (HD) planting programme (200 palm/ha) compared to 148 palm/ha (normal-density planting), owing to its short rachis length. If the trial had implemented HD for P126, the hypothetical data would have been like these, FFB 32.7 t/ha/yr and OY 8.86 t/ha/yr.

(d) *FELDA Sungai Tekam, Pahang*

FELDA has conducted a trial with D×P standard planting materials obtained from various plantation agencies to be compared with two MPOB clones. A trial plot was planted with D×P from nine agencies (Agency A-I) and two MPOB clones in 1999. Table 5.5 shows the yield record at fifth year after planting. Clone P456 showed the highest FFB of 34.48 t/ha/yr, while P126 showed the highest O/B of 30.18%. If the trial had implemented HD for P126, the hypothetical FFB and OY would be 35.7 t/ha/yr and 10.8 t/ha/yr, respectively.

Table 5.5 Performance of MPOB clones against agencies' D×P at the fifth year after planting in Sungai Tekam, Pahang, Malaysia

Agency	FFB t/ha/yr	O/B (%)	OY (t/ha/yr)
MPOB P126	26.43	30.18	7.28
MPOB P456	34.48	29.05	9.14
Agency A	28.16	29.68	7.62
Agency B	27.27	28.38	7.06
Agency C	26.85	28.42	6.96
Agency D	28.26	27.41	7.06
Agency E	25.66	26.49	6.20
Agency F	24.96	25.70	5.85
Agency G	28.71	25.86	6.77
Agency H	27.81	27.61	7.01
Agency I	25.58	27.19	6.34

5.4 MPOB Superior Clones for High-Yield

5.4.1 Clone P456

P456 is an outstanding MPOB standard clone because of its high OY, very low mantling rate and ease in tissue culturing. Recent planting of 20,000 P456 ramets in Sabah by a private company showed that the mantling rate is about 0.013% (Zamzuri 2011). Figure 5.3 shows mature fruit bunches and cross-section of fruits of clone P456. This clone was officially commercialized as Clonal Palm Series 1 (CPS1) through Malaysian Commercialization Year 2017.

5.4.2 Clone P126

Clone P126 features a distinctive characteristic, which is short rachis length (RL) of approximately 4.5 m as compared to 5–6 m of D×P and other clones. The short RL of this clone makes it suitable for HD planting (198 palms/ha) as compared to 148 palms/ha for normal density. P126 also produces smaller bunches, which is favourable for ease of harvesting. Despite smaller bunches, the expected annual FFB yield for 1 ha of P126 is 32.6 tonnes, which is still far more productive compared to D×P (28 t/ha/yr as stated by Azman and Mohd Noor 2002). In addition, clone P126 showed no incidence of mantling when field tested at various locations

Fig. 5.3 Palm, bunches and cross-section of P456 fruits

Table 5.6 Performance of P126 in high density planting design

Age (years)	Age of clones (years)					
	4	5	6	7	8	9
FFB/palm (kg)	95.72	133.21	167.52	175.86	160.30	164.42
Bunch no.	28.8	29.84	26.92	24.23	22.11	17.81
Bunch weight (kg)	3.36	4.44	6.22	7.25	7.19	9.20

Fig. 5.4 Palms, bunches and cross-section of clone P126 fruits

(Zamzuri 2011). The yield data are presented in Table 5.6. Figure 5.4 shows photos of P126 clone.

5.4.3 Clone P379

A high-yielding palm is usually related to high production of female bunches, i.e. having a high sex ratio of female to male inflorescences. From preliminary observations on some trial plots, clone P379 exhibited the characteristic of having frequent male inflorescences (MI) along with fruit bunches especially during dry spell (Zamzuri 2011). P379 (Fig. 5.5) also has potential to be incorporated with other clones as pollen source palms, substituting the use of D×P palms. The incorporation of D×P palms in clonal plots, planted in certain ratios such as 1:4, is to ensure sufficient pollen supply for pollinating clonal female inflorescences. Despite their usefulness, planters are concerned that this may cause a mix-up in the harvested fruit bunches that will create inconsistency in yield performance. Thus, planting clone P379 which potentially acts as source of pollen for pollination could produce consistent

Fig. 5.5 Palm, bunches and cross-section of P379 fruits

fruit bunches. The performance of P379 in a clonal trial at Tawau, Sabah, is as follows: FFB 182.69 kg (2010–2016), O/B of 32.87%, BNO 13.66 and BWT 13.3 kg.

5.4.4 Clone P325

Another potential clone is P325 (Fig. 5.6), which exhibited zero mantling was planted in a trial plot in Johor (Zamzuri 2011). Latest result showed that the mean FFB production was 195.7 kg/p/yr from clones ages between six and nine years with a very high O/B of 37.5%, giving an estimated OY of 10.86 t/ha/yr.

5.5 Issues and Challenges

5.5.1 Availability of High-Quality Ortets

The availability of high-quality ortets which originates from a good and proper breeding programme is prerequisite for a successful cloning process. This programme enables the identification of superior mother palms. A breeding programme also demands large areas of good land for the purpose of producing elite ortets. For a large-scale propagation of oil palm clones, a large number of elite ortets would be required. The Malaysian Standard (MS) on “Oil Palm Ortet Selection for Cloning–Specification” was developed to check production authenticity for desired productivity and sustenance of the oil palm industry, whereby only high-quality ramets, derived from high-quality ortets, are produced. Details of the MS specification are as mentioned by Zamzuri (2011).

However, the number of quality ortets that met the MS criteria is very limited. Based on selection in progeny trials, only ~2–11% of the palms are suitable as ortets (Kushairi et al. 2006). One of the alternatives that could ensure the availability of elite ortets is through recloning of the proven clonal palms as clonal ortets. Besides that, production of clonal seeds can also be considered as an alternative. This could be exploited by creating clonal parents with good combining ability.

5.5.2 Low Somatic Embryogenesis Rate

Somatic embryogenesis is a progressive development of a plant somatic cell dedifferentiated to a totipotent embryonic stem cell possessing the ability to develop to an embryo under optimal conditions. This new embryo can further evolve into a complete plant (Guan et al. 2016). To date, the low rate of embryogenesis of 3–6% (Rajanaidu et al. 1997; Corley and Tinker 2003) remains the stumbling block to large-scale ramet production. One of our efforts at MPOB is to improve embryogenesis rate. Recently, improved rates of 8.2 and 29.3% from several elite *tenera* and clonal *tenera* ortets were reported (Siti Rahmah et al. 2017).

Fig. 5.6 Palm, bunches and cross-section of P325 fruits

5.5.3 Inefficiency in Current Tissue Culture Process

The solid culture system is regarded as the "first generation" technique, which is characterized by high input of manual labour and low level of automation. This process is laborious and a high cost of production especially when scaling up. It was for this reason that led to a concentration of efforts to further improve the tissue culture process by developing plant propagation methods in liquid media. For mass propagation, the liquid culture system is one option to address the high input. At MPOB, the liquid culture system was developed to overcome the proliferation rate of friable embryogenic calli. The system basically manipulates shaking or stirring activity so that the cultures could be exposed efficiently to oxygen and nutrients as compared to the gelled system with only partial exposure.

5.5.3.1 Oil Palm Liquid Culture–MPOB Protocol

MPOB has developed the basic protocol for a liquid culture system (Tarmizi 2002) using shake flasks. It includes selection of suitable callus (friable type), media formulation, sieving of specific aggregate sizes, maturation induction, embryoid regeneration and production of rooted plantlets which involve a dual phase system (solid and liquid media). About 1000 normal-looking shoots could be obtained from one gram of mature aggregates from prolific embryogenic clones after 9–12 months in solid medium. Several innovations were developed to enhance the liquid culture system.

5.5.3.2 MPOB Fast Transfer Technique (MoFaTT) in Liquid Culture

In a shake flask culture system, the operators, in general, have to move the flasks out of the culture rooms to the laminar flow chambers to replenish the nutrient medium for culture growth sustenance. A system (MoFaTT) was developed as a rapid and convenient means for liquid medium replenishment without having to move the cultures to and fro from the culture rooms to the sterile cabinets (Tarmizi and Zaiton 2005). The benefits of the MoFaTT system (Fig. 5.7) are that the conventional ten-step protocol could be reduced to two steps for medium replenishment. In fact, the medium replenishment could be undertaken on the shaker itself at any time. The risk of contamination would thus be reduced.

5.5.3.3 Two in One MPOB Simple Impeller (2-in-1 MoSLIM) in Liquid Culture System

In a conventional process, the establishment, maintenance and maturation of liquid cultures were limited to flask and shaker sizes (Tarmizi 2002). For scaling up of cultures, bioreactors or special commercial flasks were used, but these were expensive (Tarmizi et al. 2003). Furthermore, most systems use different devices for agitation and aeration. To address this issue, the two-

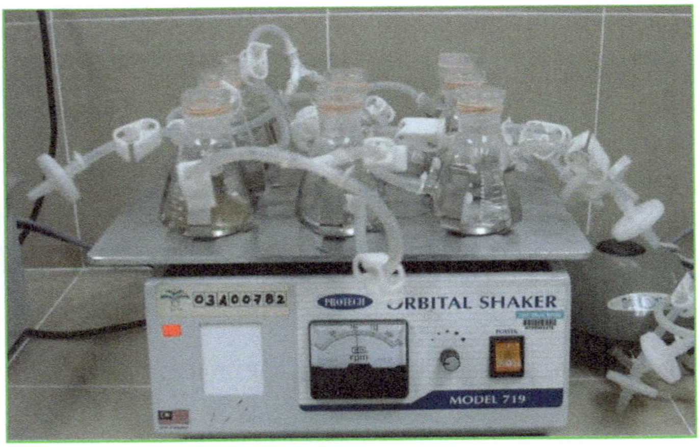

Fig. 5.7 MoFaTT system

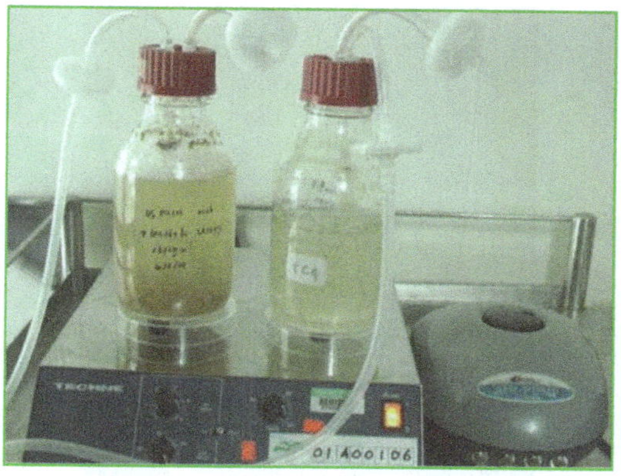

Fig. 5.8 MoSLIM system

in-one MoSLIM (Fig. 5.8) was developed as a new innovation process to provide simultaneous aeration and agitation (two in one) for the establishment, maintenance and maturation of liquid the cultures (Tarmizi and Zaiton 2006a). This approach is economical and practical for culture maintenance in liquid culture medium. The two-in-one MoSLIM is able to increase fresh weight of cell aggregates by two–six-fold from five oil palm clones after multiplication for 30–40 days.

5.5.3.4 Simple Impeller with Fast Transfer Technique (SLIM–FaTT) in Liquid Culture System

In the two-in-one MoSLIM, there is still a need to conduct replenishment of media in a laminar flow cabinet. Thus, there is still a risk of contamination. To overcome this problem, the system was further modified so that no movement of cultures is required, and medium replenishment could be carried out at the culture site itself. The new system is a combination of the two earlier innovations (MoFaTT and MoSLIM) and is called SLIM–FaTT (Fig. 5.9). The added benefit of the SLIM–FaTT system is that the conventional eight-step protocol is reduced to one step for media replenishment. The benefits from both of the earlier systems also apply in this new system as well (Tarmizi and Zaiton 2006b).

5.5.3.5 Multiplication of Oil Palm Liquid Cultures in Bioreactors

A bioreactor technology was primarily developed for biomass production. Hence, the culture vessels needed adaptations to the specific requirements of embryogenic suspension. The purpose of using a bioreactor is for large-scale propagation. From the previous studies at MPOB, the fresh weight of the cultures from selected clones increased 4–20-fold after 50–80 days in a B Braun Bench-top two litre Bioreactor and five-fold after about 60 days in a Biotron bioreactor (Fig. 5.10). Figure 5.11 shows the distinctive growth increase of cultures in a bioreactor as compared to the shake flask system.

The present findings on the bioreactor system show the possibility of large-scale production of embryogenic suspension cultures in a single run. Once the bioreactor method has become more efficient, time and space could be saved.

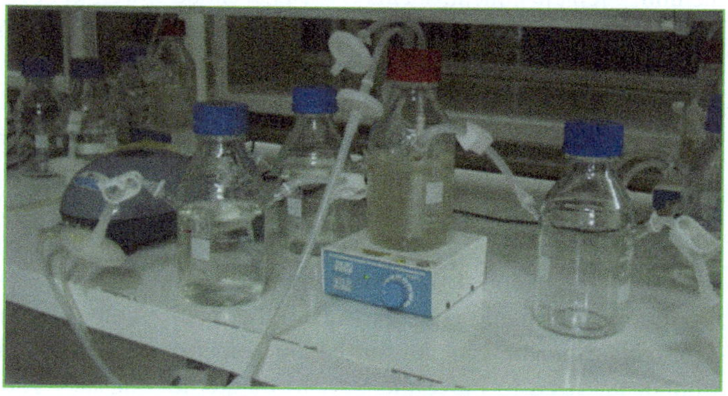

Fig. 5.9 SLIM-FaTT system

Fig. 5.10 Proliferation in a bioreactor

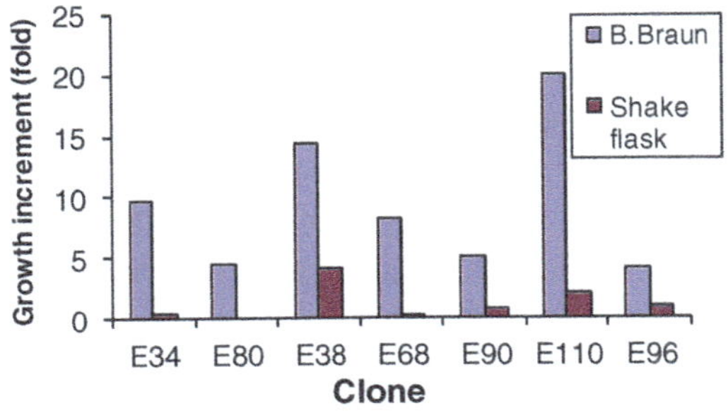

Fig. 5.11 Growth comparison of oil palm cultures maintained in a B.Braun bioreactor or shake flask after 50 days incubation

5.5.3.6 MPOB Modified Vessel (MoVess) for Liquid Tissue Culture System

This innovation is a vessel modified with adaptors for controlling growth conditions such as medium replenishment. This vessel, called MoVess (Fig. 5.12), was designed to overcome the high costs of a bioreactor and at the same time provides a simpler culture inoculation operation (Tarmizi et al. 2007). Fresh weight of cell aggregates increments of up to 35-fold were obtained for cultures of selected oil palm clones after 30–60 days growth in MoVess. This new system is simple and more economical for scaling up oil palm liquid cultures. Again, the medium replenishment can be carried out on site without the need to move the vessel to a sterile cabinet, which also reduces the risk of contamination.

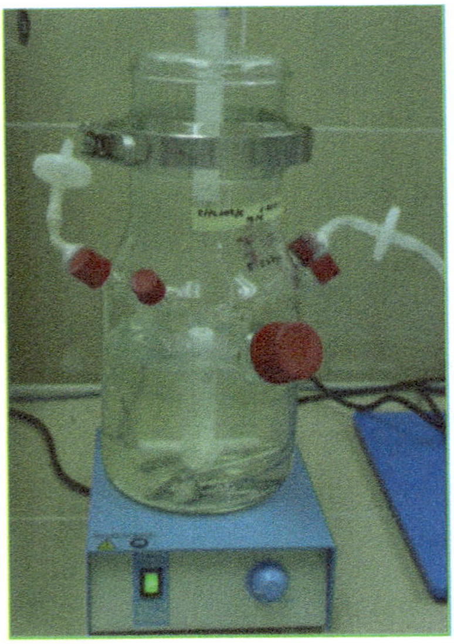

Fig. 5.12 MoVess system

5.5.3.7 MPOB Motorized Vessel (MPOB-Motovess) for Liquid Tissue Culture

Motovess is an improved version of MoVess (Tarmizi et al. 2007). This system omits the use of the magnetic stirrer used in MoVess. The magnetic stirrer limits the culture volume between 1 and 2 L, and omission of the stirrer will allow handling of a larger culture volume of between 2 and 9 L (Fig. 5.13). MotoVess consists of a motor with a stand and a shaft with impeller, made of perforated stainless steel for agitation and aeration of the medium. It was recorded that two–six-fold increment in fresh weight of cultures could be obtained after 40 days using MotoVess.

5.5.3.8 Motorized Vessel with Fast Media Transfer (MoVeFast)

MPOB motorized vessel (5–10 L capacity) was developed to propagate oil palm cultures for increased production (Tarmizi et al. 2009). The device offers economical and practical means of culturing cell aggregates in larger volumes of liquid media compared to traditional shake flasks. The system is not fully automated, whereby the vessel needs to be brought to a laminar flow cabinet for media replenishment in

Fig. 5.13 MotoVess system

a sterile manner. However, handling of the system is inconvenient and impractical owing to bulkiness and weightiness of the vessel. The risk of contamination also increases by having to substitute used media with fresh media. To overcome these problems, a mechanism was developed to integrate the media replenishment process. This new system is called MoVeFast (Fig. 5.14).

In a MoVeFast system, a stainless steel scaffold with a raised platform holding a bottle containing at least five litres of fresh media is connected to a motorized vessel. To ensure efficient media replenishment, the top of the bottles is equipped with top plates attached to vent devices. Clamps attached to tubings are used for controlling the flow of media during media replenishment. On the other end of the motorized vessel, an empty pre-sterilized modified bottle of preferred size, placed on the lower level of the scaffold, is connected for the purpose of discarding spent media. This new system allows media replenishment on site. This system generated a fresh weight increment of about five-fold for cultures of selected oil palm clones after about 40 days in the MoVeFast system.

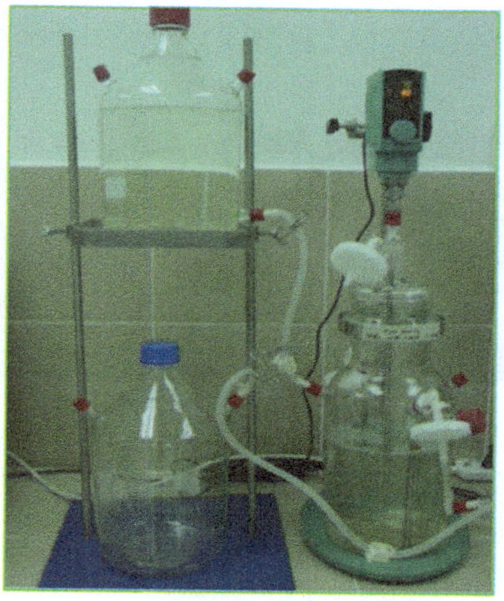

Fig. 5.14 MoVeFast system

5.5.3.9 MultiVessel (MV) Bioreactor for Liquid Tissue Culture System

Improvements to the oil palm liquid culture system are a continuous process. This led to the development of an innovative technology, namely the MultiVessel (MV) bioreactor (Fig. 5.15), for the simultaneous multiplication of cell aggregates of various clones and/or application of various treatments (Tarmizi et al. 2016). MV provides convenient alternative to the conventional shake flask system. Multiplication of cultures in the MV bioreactor does not require any shaker or a large area. With a working volume of 300–700 ml, this system uses a simple impeller and a pump for agitation and aeration purposes. Basically, the MV bioreactor is an improvement of the MPOB Simple Impeller (two-in-one MoSLIM) system, previously developed using commonly available Schott bottles (Tarmizi and Zaiton 2006a). However, the two-in-one MoSLIM system can only multiply cultures of a single clone or for the application of a single treatment. To overcome this problem, improvements were made to the system to enable more vessels to be connected to single pump(s). Moreover, this system is more cost effective than a commercial lab scale multi-

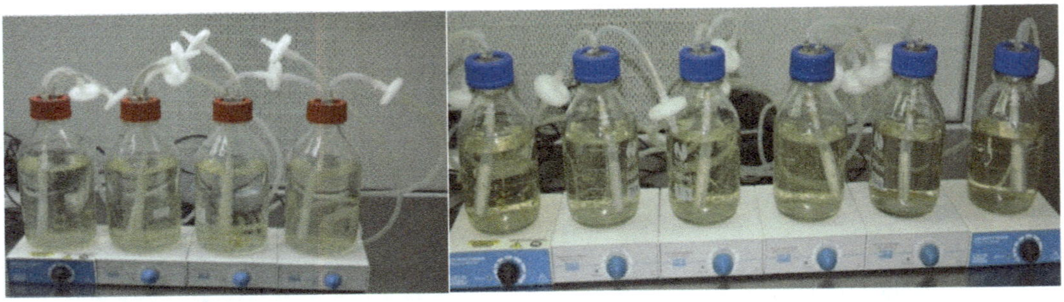

Fig. 5.15 MultiVessel (MV) 4 and 6 systems

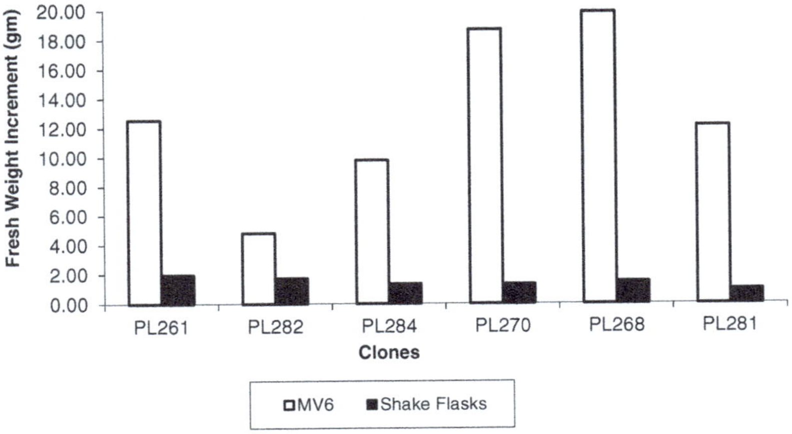

Fig. 5.16 Fresh weight increment of cultures for clones after approximately 30 days of multiplication in MV6, compared to the shake flask system (conventional method)

fermenter, whereby the preparation for culture inoculation is also tedious. This new system can be applied to liquid culture systems of any crop with further potential for automation. Two to 14-fold increments in fresh weight of cultures were obtained after about 30 days (Fig. 5.16) for oil palm clones when multiplied in MV4 (four vessels) and MV6 (six vessels) systems. Normal regeneration of cell aggregates was observed (Fig. 5.17).

The application of liquid media technique has opened up possibilities for automation of the system. Propagation via liquid media has also proven to increase the number of embryogenic cultures by several folds. This is due to the nature of the system in which the cell cultures are fully surrounded by media hence allowing efficient nutrient transfer into cells and removal of metabolic waste out of these cells. The rapid development of suspension cultures showed that a reliable alternative method is available for multiplication of friable embryogenic calli in the oil palm (Soh et al. 2011; Tan et al. 2003; Tarmizi and Zaiton 2005). However, as the conventional culture system does not permit the regeneration of these cultures or embryoids directly in liquid system, the best option is to synergize the use of both, solid and liquid cultures (shake flask and bioreactor).

5.6 Land Scarcity

In a report by Khoo and Chandramohan (2002), the authors indicated that prime agriculture land is currently facing serious shortage. Further expansion of land would risk coming under fire by the environmental non-governmental organizations. Moving to or using marginal areas with poorer soils, terrain and rainfall as an alternative would inevitably lead to escalating problems. Planting in these areas would cost more to

Fig. 5.17 **a** Cell aggregates of clone PL261 from the MV6 system, **b** regeneration of cell aggregates and **c** development into embryoids and shoots

develop, is less productive which would also lead to higher production costs. In addition to that, when production figures are tabulated, it would eventually affect the overall national oil yield!

Taking all these into consideration, it is obvious that to achieve the increase in production output by expanding land areas devoted to oil palm is not an option. Therefore, that leaves the fact that yield production will have to be increased substantially within the same amount of land planted with the crop. This is where quality planting materials such as high-yielding clones play a pivotal role in improving the national productivity.

5.7 Supply and Demand of Clones

The current production of clonal oil palm is still low when compared to the demand for this material. The total production of ramets by major tissue culture laboratories in the country is approximately 5 million (MPOB, unpublished data). To improve this situation, each company that is involved in clonal production should be encouraged to increase its production capacity to more than half million ramets annually.

For this purpose, MPOB strongly encourages big and medium-size plantations to set up their own tissue culture laboratories. To help them in this endeavour, MPOB is providing consultations and training. Up till now, MPOB had provided consultations and training to seven commercial tissue culture laboratories. However, it should be noted that one of the most important requirements for a successful venture in the production of oil palm clonal materials is the availability of high-quality ortets, selected from a comprehensive breeding programme. Such programme is essential for the identification of superior mother palms. Thus, tissue culture method is not a standalone field. The current listing of tissue culture laboratories worldwide is as follows:

Malaysia:

1. FELDA Agricultural Services Sdn Bhd
2. Applied Agricultural Resources Sdn Bhd (AAR)
3. IOI Corporation Bhd
4. Sime Darby Plantation Sdn Bhd
5. United Plantations Bhd
6. Clonal Palms Sdn Bhd
7. Agrocom Enterprise Sdn Bhd
8. TSH Biotech Sdn Bhd.
9. Sawit Kinabalu Bio-Tech Sdn Bhd
10. Sabah Softwoods Berhad
11. KULIM-Top Plant Sdn Bhd
12. RISDA
13. FELCRA Bhd
14. Tradewinds Plantation Bhd
15. MPOB.

Indonesia

1. Indonesian Oil Palm Research Institute (IOPRI)
2. PT. Socfin Indonesia
3. SUMBIO (PT. London Sumatera/IndoAgri)
4. PT. Bina Sawit Makmur (Sampoerna Agro)

5. PT. Tunggal Yunus Estate (Asian Agri)
6. PT. Dami Mas Sejahtera (Sinar Mas)
7. PT. Bakti Tani Nusantara
8. Sarana Inti Pratama (IndoAgri)
9. Bakrie-ASD (Bakrie)
10. Sasaran Ehsan Mekarsari
11. *Dura* Inti Lestari (Darmex Agro).

Costa Rica
1. Agricultural Services & Development (ASD).

Papua New Guinea
1. Dami Oil Palm Research Station Biotechnology Laboratory, New Britain Palm Oil Ltd.

Thailand
1. Univanich Palm Oil Public Company Ltd.

Higher production of clonal palms from both existing and new laboratories will definitely reduce the cost per ramet. The current price for a bare-rooted ramet ranges from USD 5–10 per ramet (MPOB 2006). At a reduced price, more growers would plant tissue culture materials in their plantations. It is envisaged that this will increase Malaysia's crude palm oil (CPO) production annually.

Currently, the estimated land planted with clonal materials in Malaysia is about 200,000 hectares (MPOB, unpublished data). Based on the age profile of the Malaysian oil palm, approximately 6.4% (365,581 ha) of the total oil palm planted area is represented by old palms, which are more than 25-year old in 2016 (Shahari and Ismail 2019). Based on this figure, the national requirement of new planting materials is approximately 54.1 million palms. With the current production capacity of 5 million ramets per year, oil palm companies and smallholders can incorporate 10% of clonal materials into their replanting programmes.

Independent smallholders contribute about 16.3% (934,000 ha) of the total hectarage planted with oil palm. To realize the national target of high-quality planting materials, MPOB will initiate a programme called "Contract farming", specifically targeted for independent smallholders. In this concept, an identified company would purchase FFB from the smallholders at a premium price, based on higher expected OER of clones compared to the D×P. The clones will be planted in areas near the palm oil mills. The company, jointly with MPOB, will also provide advisory services on good agricultural practices (GAP) to smallholders.

5.8 Standard for Licensing Policies

Commercialization of clonal materials requires specific standards and license in order to protect buyers from counterfeit materials. Companies are required to meet specific criteria prior to license issuance. In Malaysia, licenses are issued for three activities, namely producing plants from oil palm tissues, selling and moving plants from oil palm tissues and storing plants from oil palm tissues. The fee for each activity is RM100.00 per year.

5.9 Recent Advances in Oil Palm Tissue Culture

5.9.1 Molecular Markers for Quality Control

Research groups working on the oil palm have reported numerous gene expression studies for both embryogenesis and abnormalities arising from tissue culture (Low et al. 2008; Ho et al. 2007; Adam et al. 2007a; Lin et al. 2009; Roowi et al. 2010; Beule et al. 2011; Habib et al. 2014). These studies mainly strive to discover possible biomarkers to address the issues of tissue culture amenity and abnormalities. In the medical field, biomarkers were developed for the purpose of predicting disease susceptibility, progression and drug response; hence the term personalized medicine was coined (Hood et al. 2004; Jain 2004; Nevins et al. 2003). However, in the case of the oil palm, the availability of a predictive tool for embryogenesis and abnormality would be prioritized for screening and perhaps leading to improvements in the process to prevent the undesirable conditions.

Two main issues affecting oil palm clonal propagation are the poor embryogenesis rate and the risk of obtaining the mantled fruits. The latter is discussed in the next section. The average embryogenesis rate has not improved from the previously reported 6% (Wooi 1995). As oil palm tissue culture goes through somatic embryogenesis, callusing, embryogenesis and shoot regeneration are the main factors crucial to the success of clonal propagation. Callusing is considered not an issue for palms with rates ranging from 11 to 20% (Soh et al. 2011; Ho et al. 2009). The formation of embryogenic callus which generally leads to somatic embryo formation is a sporadic and unpredictable process. As phytohormones, primarily auxin, are used in early tissue culture, studies to investigate targeted molecular pathways involving auxin have been reported. Several gene families including the Aux/IAA gene family are rapidly induced by auxin (Abel and Theologis 1996; Hagen and Guilfoyle 2002; Nemhauser et al. 2006; Paponov et al. 2008). EgIAA9, a putative Aux/IAA gene, was moderately associated with somatic embryogenesis potential (Ooi et al. 2012). Several genes including EgHOX1, Eg707 and EgPK1 have been reported to be highly expressed in embryogenic callus compared to non-embryogenic callus (Ooi et al. 2008, 2016; Thuc et al. 2011). Ectopic expression of EgAP2-1 could enhance regeneration capacity in transgenic Arabidopsis (Morcillo et al. 2007). Other molecular approaches have also been taken to identify molecular markers for embryogenesis such as genomics (Ting et al. 2013; Tranbarger et al. 2012), epigenetics (Ho et al. 2013) and proteomics (Tan et al. 2016; De Carvalho et al. 2014).

5.9.2 Discovery of an Epigenetic Marker for the Mantled Somaclonal Variant

In the 80s, a somaclonal variant arising from oil palm clonal propagation was reported which resulted in fruit abortion and yield losses (Corley et al. 1986). This phenotype, called mantled, involves feminization of the male staminodes in female inflorescences and occasionally the stamens in the male inflorescences as well. Due to the range of severity and reversible occurrences of mildly mantled phenotypes, an epigenetics origin had been widely hypothesized (Jaligot et al. 2000). Numerous studies had been conducted to investigate epigenetic changes particularly DNA methylation, in relations to mantling (Jaligot et al. 2000; Tregear et al. 2002; Jaligot et al. 2004; Lei et al. 2006; Jaligot et al. 2014; Ong-Abdullah et al. 2015).

Floral organ identity is known to be regulated mainly by the MADS-box family of genes (Coen and Meyerowitz 1991). As the mantling abnormality involves changes in floral organs, expression studies on the oil palm B-type MADS-box genes showed that EgDEF1 and EgGLO2 genes were downregulated in developing mantled inflorescences (Adam et al. 2007a, b). Further investigations revealed no DNA methylation differences between normal and mantled clones in two retrotransposons located within the introns of EgDEF1 (Jaligot et al. 2014), but demethylation of the Karma retroelement, also located within an intron of EgDEF1, was found to be involved in defining the mantled phenotype (Ong-Abdullah et al. 2015). Subsequently, a KARMA screening assay was developed commercially that can be used to screen nursery plantlets (Ong-Abdullah et al. 2016). This would allow assessment of the mantling risk and early culling of plants to circumvent economic and time losses involved in the event of planting mantled palms.

5.10 Cloning of Oil Palms with Special Traits

Although establishing high-yielding planting materials is of priority, palms with other important economic traits have also been identified. Selected oil palm possessing agronomic and secondary traits (as below) has also been cloned:

i. High bunch index
ii. Low height
iii. High vitamin E
iv. High carotene
v. Virescens.

5.10.1 High Bunch Index

The bunch index (BI) of the present planting materials is about 0.4. The benchmark to be achieved is a bunch index of >0.6. A Tanzanian virescens *dura* (with a BI of 0.68) and a *tenera* (with BI = 0.58) were cloned. About 50 ramets of both clones are already planted in the field.

5.10.2 Low Height Increment (Compact Palm)

The height increment of the present planting materials ranged from 45 to 75 cm/year. The benchmark to be achieved is palm with a height increment of <30 cm/year. More than low height increment, ortets were cloned, and the cultures are in various forms of growth, from callus to nursery stages. The first batch of 21 ramets cloned from selected low height increment palms was field planting for further evaluation.

5.10.3 High Vitamin E

The current vitamin E content of the present planting materials is about 600 ppm. The benchmark to be achieved is vitamin E content ranging from 1000 to 1500 ppm (Mohd Din et al. 2005). A *dura* palm (with 1551 ppm of vitamin E) and a *tenera* (containing 1392 ppm vitamin E) were cloned. A limited number of ramets are already planted for further observation.

5.10.4 High Carotene

The current carotene content of *E. guineensis* is 500 ppm, whereas for *E. oleifera*, it is 1500 ppm. The benchmark to be achieved for the fast track programme is 2000 ppm for *E. guineensis* and 3000 ppm for *E. oleifera*. The oil from *E. oleifera* is advocated as potentially useful for the neutraceutical industry (Choo and Yusof 1996). Out of 22 *E. oleifera* ortets sampled for cloning using various types of culture media, only one palm with 3146 ppm carotene content was successfully cloned (Zamzuri and Siti Rahmah 2007). A total of 53 ramets were produced and planted in Hulu Paka, Terengganu, Malaysia.

5.10.5 Virescens (Clone PS1, P505)

The origin of this clone is from cross-breeding of Nigerian *dura* and AVROS *pisifera*. This clone produced *virescens* fruit. *Virescens* fruits are green when unripe and change to orange when the bunch matures (Fig. 5.18) (Singh et al. 2014). Since *virescens* fruits undergo a more profound colour change upon ripening, it is easier to identify ripe bunches, particularly in tall palms where they can be obscured by fronds, thus minimizing yield loss due to fallen fruits or harvesting of unripe bunches. In terms of yield, the ortet produces an average of 28.9 t/ha/yr of FFB, 30.14% O/B and OY of 8.71 t/ha/yr.

Fig. 5.18 Virescens palms are easily monitored, immature bunches are green, and matured bunches are orange in colors

5.11 Conclusion

Oil palm improvement via conventional breeding is a slow process. Cloning is an alternative method to propagate true-to-type genotypes with desired traits. Cloning also provides means to exploit exotic genetic materials such as palms with high carotene and high-unsaturated oil contents. Planting clones can be the way forward to increase the oil palm productivity, and thus, the industry should actively include clonal materials for replanting programme. Initially, in vitro production of oil palm clones on a large-scale was hampered by the abnormality problem. However, the in vitro cloning process of oil palm has tremendously improved since its early days, and now, the chances of planting mantled palms can be minimized with the discovery of KARMA gene. Based on this discovery, a diagnostic assay, *SureSawit*^{TM} KARMA, has been developed for early screening of clonal plantlets and prediction of mantled fruit.

Field results of clonal materials have been very encouraging. The superior performance of clones is due to the inheritance of good traits. Although environmental factors such as the soil type can affect yield, the performance of clones thus far was better than the standard D×P. With reliable ortet selection, it is also possible to forecast the clonal potential based on the heritable traits applied in ortet selection. As clonal materials are generated from elite high-yielding palms, these materials would help to boost the overall oil yield and create the "second wave" in yield improvement.

References

Abel S, Theologis A (1996) Early genes and auxin action. Plant Physiol 111:9–17

Adam H, Jouannic S, Orieux Y, Morcillo F, Richaud F, Duval Y, Tregear JW (2007a) Functional characterization of MADS box genes involved in the determination of oil palm flower structure. J Exp Bot 58:1245–1259

Adam H, Jouannic S, Morcillo F, Verdeil JL, Duval Y, Tregear JW (2007b) Determination of flower structure in *Elaeis guineensis*: do palms use the same homeotic genes as other species? Ann Bot 100:1–12

Azman I, Mohd Noor M (2002) The optimal age of oil palm replanting. Oil Palm Industry Econ J 2(1):11–18

Basiron Y (2007) Palm oil production through sustainable plantations. Eur J Lipid Sci Technol 109:289–295

Beule T, Camps C, Debiesse S, Tranchant C, Dussert S, Sabau X, Jaligot E, Alwee SSRS, Tregear JW (2011) Transcriptome analysis reveals differentially expressed genes associated with the mantled homeotic flowering abnormality in oil palm (*Elaeis guineensis*). Tree Genet Genomes 7:169–182

Choo YM, Yusof B (1996) *Elaeis oleifera* palm for the pharmaceutical industry. PORIM Information Series No. 22: 4 p

Coen ES, Meyerowitz EM (1991) The war of the whorls: genetic interactions controlling flower development. Nature 353:31–37

Corley RHV, Tinker PB (2003) The oil palm, 4th edn. Blackwell Science Ltd., Oxford (Monograph of growth, botany and use of oil palm)

Corley RHV, Lee CH, Law IH, Wong CY (1986) Abnormal flower development in oil palm clones. Planter 62:233–240

De Carvalho Silva R, Carmo LST, Luis ZG, Silva LP, Scherwinski-Pereira JE, Mehta A (2014) Proteomic identification of differentially expressed proteins during the acquisition of somatic embryogenesis in oil palm (*Elaeis guineensis* Jacq.). J Proteomics 104:112–127

Guan Y, Li SG, Fan XF, Su ZH (2016) Application of Somatic Embryogenesis in Woody Plants. Front Plant Sci 7:938

Habib SH, Ho CL, Syed Alwee SSR, Namasivayam P (2014) Molecular analysis on the shoot apical meristem of truncated leaf syndrome plantlets of oil palm (*E. guineensis* Jacq.) Plant Cell Tissue Organ Cult 120:1023–1036

Hagen G, Guilfoyle T (2002) Auxin-responsive gene expression: genes, promoters and regulatory factors. Plant Mol Biol 49:373–385

Hashim AT, Ishak Z, Rosli SK, Ong-Abdullah M, Ooi SE, Husri MN, Bakar DA (2018) Oil palm (*Elaeis guineensis* Jacq.) somatic embryogenesis. In step wise protocals for somatic embryogenesis of important woody plans. Springer, Cham, pp 209–229

Ho CL, Kwan YY, Choi MC, Tee SS, Ng WH, Lim KA, Lee YP, Ooi SE, Lee WW, Tee JM, Tan SH, Kulaveerasingam H, Syed Alwee SSR, Ong-Abdullah M (2007) Analysis and functional annotation of expressed sequence tags (ESTs) from multiple tissues of oil palm (*Elaeis guineensis* Jacq.). BMC Genom 8:381

Ho YW, Tan CC, Soh AC, Wong G, Chong SP, Choo CN, Norazura A (2009) Biotechnological approaches in producing oil palm planting material—a success story. Int J Oil Palm 6:86–93

Ho WK, Ooi SE, Mayes S, Namasivayam P, Ong-Abdullah M, Chin CF (2013) Methylation levels of a novel genetic element, EgNB3 as a candidate biomarker associated with the embryogenic competency of oil palm. Tree Genet Genom 9(4):1099–1107

Hood L, Heath JR, Phelps ME, Lin B (2004) Systems biology and new technologies enable predictive and preventative medicine. Science 306(5696):640–643

Jain KK (2004) Role of pharmacoproteomics in the development of personalized medicine. Pharmacogenomics J 5(3):331–336

Jaligot E, Rival A, Beule T, Dussert S, Verdeil JL (2000) Somaclonal variation in oil palm (*Elaeis guineensis* Jacq.): the DNA methylation hypothesis. Plant Cell Rep 19:684–690

Jaligot E, Beule T, Baurens FC, Billotte N, Rival A (2004) Search for methylation-sensitive amplification polymorphisms associated with the 'mantled' variant phenotype in oil palm (*Elaeis guineensis* Jacq.). Genome 47:224–228

Jaligot E, Hooi WY, Debladis E, Richaud F, Beule T, Collin M, Agbessi MDT, Sabot F, Garsmeur O, D'hont A, Alwee SSRS, Rival A (2014) DNA methylation and expression of the EgDEF1 gene and neighboring retrotransposons in mantled somaclonal variants of oil palm. PLoS ONE 9:e91896

Jones LH (1974) Propagation of clonal palms by tissue culture. Oil Palm News 17:1–8

Khaw CH, Ng SK (1997) Performance of commercial scale clonal oil palm (*Elaeis guineensis* Jacq.) plantings in Malaysia. Paper presented at international symposium on biotechnology of tropical and subtropical species, Brisbane 29 September 1997

Khoo KM, Chandramohan D (2002) Malaysian palm oil industry at crossroads and its future direction. Oil Palm Industry Econ J 2(2):10–15

Kushairi A, Tarmizi AH, Zamzuri I, Ong-Abdullah M, Rohani O, Samsul Kamal R, Ooi SE, Ravigadevi S, Mohd Basri W (2006) Current status of oil palm tissue culture in Malaysia. In: Kushairi A, Sambanthamurthi R, Ong-Abdullah M, Chan KC (eds) Proc. Clonal & Qty. Rep. Material. Malaysian Palm Oil Board, Malaysia, pp 3–14

Lei CP, Jiun KS, Choo CS, Singh R (2006) Analysis of tissue culture-derived regenerants using methylation sensitive AFLP. Asia Pacific J Mol Biol Biotech 14:47–55

Lin HC, Morcillo F, Dussert S, Tranchant-Dubreuil C, Tregear JW, Tranbarger TJ (2009) Transcriptome analysis during somatic embryogenesis of the tropical monocot *Elaeis guineensis*: evidence for conserved gene functions in early development. Plant Mol Biol 70:173–192

Low ETL, Alias H, Boon SH, Shariff EM, Tan CYA, Ooi LCL, Cheah SC, Raha AR, Wan KL and Singh R (2008) Oil palm (*Elaeis guineensis* Jacq.) tissue culture ESTs: Identifying genes associated with callogenesis and embryogenesis. BMC Plant Biol 8:62

Maheran AB, Abu Zarin O, Aw KT and Chin CW (1995) FELDA's early experiences with vegetative propagation of the oil palm (*Elaeis guinensis* Jacq.). In: Jalani S, Ariffin D, Rajanaidu N, Tayeb MD, Paranjothy K, Basri MW, Henson IE, Chang KC (eds) Proceedings of the 1993 PORIM international palm oil congress on update and vision. PORIM, Kuala Lumpur. pp 99–113

Mohd Din A, Rajanaidu N, Kushairi A (2005) Exploitation of genetic variability in Oil Palm. In: Proceedings of MOSTA best practices workshops: agronomy and crops management. The Malaysian Oil Scientists' and Technologists' Association, pp 19–42

Morcillo F, Gallard A, Pillot M, Jouannic S, Aberlenc-Bertossi F, Collin M, Verdeil JL, Tregear JW (2007) EgAP2-1, an AINTEGUMENTA-like (AIL) gene expressed in meristematic and proliferating tissues of embryos in oil palm. Planta 226:1353–1362

MPOB (2006) National tissue culture development survey. MPOB internal report. Ministry of Primary Industry, Malaysia

Nemhauser JL, Hong F, Chory J (2006) Different plant hormones regulate similar processes through largely nonoverlapping transcriptional responses. Cell 126:467–475

Nevins JR, Huang ES, Dressman H, Pittman J, Huang AT, West M (2003) Towards integrated clinico-genomic models for personalized medicine: combining gene expression signatures and clinical factors in breast cancer outcomes prediction. Hum Mol Genet 12 (spec no 2):R153–R157

Ong-Abdullah M, Ordway JM, Jiang N, Ooi SE, Kok SY, Sarpan N, Azimi N, Hashim AT, Ishak Z, Rosli SK, Malike FA, Abu Bakar NZ, Marjuni M, Abdullah N, Yaakub Z, Amiruddin MD, Nookiah R, Singh R, Low ETL, Chan KL, Azizi N, Smith SW, Bacher B, Budiman MA, Van Brunt A, Wischmeyer C, Beil M, Hogan M, Lakey N, Lim CC, Arulandoo X, Wong CK, Choo CN, Wong WC, Kwan YY, Syed Alwee SSR, Sambanthamurthi R, Martienssen RA (2015) Loss of Karma transposon methylation underlies the mantled somaclonal variant of oil palm. Nature 525:533–537

Ong-Abdullah M, Ordway JM, Jiang N, Ooi SE, Mokri A, Kok SY, Sarpan N, Azimi N, Hashim AT, Ishak Z, Rosli SK, Nookiah R, Singh R, Low ETL, Sachdeva M, Smith SW, Lakey N, Martienssen RA, Sambanthamurthi R (2016) Tissue culture and epigenetics. Planter 92:741–749

Ooi SE, Harikrishna K, Ong-Abdullah M (2008) Isolation and Characterization of a putative serine/threonine kinase expressed during oil palm tissue culture. Special issue on Malaysia-MIT biotechnology partnership programme. J Oil Palm Res 1:14–22

Ooi SE, Choo CN, Ishak Z, Ong-Abdullah M (2012) A candidate auxin-responsive expression marker gene, EgIAA9, for somatic embryogenesis in oil palm (*Elaeis guineensis* Jacq.). Plant Cell Tissue Organ Cult 110:201–212

Ooi SE, Ramli Z, Syed Alwee SSR, Kulaveerasingam H, Ong-Abdullah M (2016) EgHOX1, a HD-Zip II gene, is highly expressed during early oil palm (*Elaeis guineensis* Jacq.) somatic embryogenesis. Plant Gene 8:16–25

Paponov IA, Paponov M, Teale W, Menges M, Chakrabortee S, Murray JAH, Palme K (2008)

Comprehensive transcriptome analysis of auxin responses in Arabidopsis. Mol Plant 1:321–337

MPOB Pocketbook (2017) Oil palm statistics. Malaysian Palm Oil Board (MPOB), Selangor, Malaysia

Rabéchault H, Martin JP (1976) Multiplication vegetative du palmier a huile *(Elaeis guineensis* Jacq.) al'aide de cultures de tissus foliares. CR Acad Sci Paris 283:1735–1737

Rajanaidu N, Jalani BS (1999) Worldwide production, performance and issues related to oil palm planting materials. In: Proceedings of the 1996 seminar on sourcing of oil palm planting materials for local and overseas joint ventures, PORIM, Bangi, pp 28–70

Rajanaidu N, Rohani O, Jalani S (1997) Oil Palm clones: current status and prospects for commercial production. Planter 73(853):163–184

Rohani O, Sharifah SA, Mohd Rafii Y, Ong M, Tarmizi AH, Zamzuri I (2000) Tissue culture of oil palm—Chapter 7. In: Yusof B, Jalani BS, Chan KW (eds) Advances in oil palm research, vol 1. Bangi, MPOB, pp 238–283

Roowi SH, Ho CL, Alwee SSRS, Ong-Abdullah M, Napis S (2010) Isolation and characterization of differentially expressed transcripts from the suspension cells of oil palm (*Elaeis guineensis* Jacq.) in response to different concentration of auxins. Mol Biotechnol 1:1–19

Shahari DN, Ismail A (2019) Overview of oil palm replanting in Malaysia. In: Ahmad kushairi D, Balu N, Ismail A (eds) Oil palm replanting: little steps to a giant leap. Bangi, MPOB, pp 27–46

Sharma M (2006) Challenges facing the Malaysian Palm Oil Industry—multi pronged strategies for raising oil yield, productivity and profitability. In: Kushairi A, Sambanthamurthi R, Ong-Abdullah M, Chan KC (eds) Proc. Clonal & Qty. Rep. Material. Malaysian Palm Oil Board, Malaysia

Simon S, Koh HL (2005) An update on performance of tissue cultured oil palm clones in PBP oil palms BHD —East Malaysia. In: Proceeding of the 2005 national seminar on practices for super high yielding plantation. Malaysian Palm Oil Board, Kuala Lumpur, p 20

Simon S, Hendry T, Chang SW, Kiaw CW (1998) Early yield performance of clonal oil palm (*Elaeis guineensis* Jacq.) plantings in PBB Oil Palm Bhd., Sabah—a case study. Planter 74(866):257–269

Singh R, Low ET, Ooi LC, Ong-Abdullah M, Nookiah R, Ting NC, Marjuni M, Chan PL, Ithnin M, Manaf MA, Nagappan J, Chan KL, Rosli R, Halim MA, Azizi N, Budiman MA, Lakey N, Bacher B, Van Brunt A, Wang C, Hogan M, He D, Macdonald JD, Smith SW, Ordway JM, Martienssen RA, Sambanthamurthi R (2014) The oil palm VIRESCENS gene controls fruit colour and encodes a R2R3-MYB. Nat Commun 5:4106

Siti Rahmah AR, Abdullah MP, Tarmizi AH, Zamzuri I, Shaharuddin NA, Che Yem MJ (2017) Evaluation on the amenability of selected oil palm (*Elaeis guineensis* Jacq.) to cloning: Elite vs. Clonal *Tenera* Ortets. Presented at International Palm Oil Congress (PIPOC), Kuala Lumpur Conventional Centre, Kuala Lumpur, 14–16 November 2017

Soh AC, Wong G, Tan CC, Chew PS, Chong SP, Ho YW, Wong CK, Choo CN, Nor Azura H, Kumar K (2011) Commercial-scale propagation and planting of elite oil palm clones: research and development towards realization. J Oil Palm Res 23:935–952

Tan CC, Wong G, Soh AC, Hor TY, Chong SP, Gopal K (2003) Experiences and lessons from oil palm clonal evaluation trials and commercial test plantings. In Proceedings of the 2003 PIPOC international palm oil congress. MPOB, Bangi, pp 1093–1119

Tan HS, Liddell S, Ong-Abdullah M, Wong WC, Chin CF (2016) Differential proteomic analysis of embryogenic lines in oil palm (*Elaeis guineensis* Jacq.). J Proteomics 143:334–345

Tarmizi AH (2002) Oil palm liquid culture—MPOB protocol: MPOB Information Series TT No. 138

Tarmizi AH, Zaiton R (2005) MPOB fast Transfer Technique (MoFaTT) in Liquid Culture System. MPOB Information Series TT No. 261

Tarmizi AH, Zaiton R (2006a) Two-in-One MPOB-Simple Impeller (2 in 1 MoSLIM) in Liquid System. MPOB Information Series TT No. 303

Tarmizi AH, Zaiton R (2006b) Simple impeller with fast transfer techniques (SLIM-FaTT) in liquid cultrure system. MPOB Information Series TT No. 304

Tarmizi AH, Norjihan MA, Zaiton R (2003) Multiplication of oil palm suspension cultures in a bench-top two litre bioreactor. J Oil Palm Res 16(2):44–49

Tarmizi AH, Zaiton R and Rosli MY (2007) MPOB Modified Vessel (MoVess) for liquid tissue culture system. MPOB Information Series TT No. 355

Tarmizi AH, Zaiton R, Rosli MY (2009) MPOB Motorized Vessel (MPOB-MotoVess) for Liquid Tissue Culture System. MPOB Information Series, TT No. 454

Tarmizi AH, Rosli MY, Zanariah R (2016) MultiVessel (MV) bioreactor for liquid culture system. MPOB Information Series, TT No, p 593

Tarmizi AH, Zamzuri I, SamsulKamal R, Ong-Abdullah M, Ooi SE, Naqiuddin MH and Dalilah AB (2018) Oil palm (*Elaeis guineensis* Jacq.) Somatic embryogenesis. In: Jain M, Gupta P (eds) Step wise protocols for somatic embryogenesis of important woody plants, vol 2. Springer International Publishing, Berlin

Thuc LV, Sarpan N, Ky H, Ooi SE, Suhaimi N, Ho CL, Ong-Abdullah M, Chin C-F, Namasivayam P (2011) A Novel Transcript of Oil Palm (*Elaeis guineensis* Jacq.), Eg707, is Specifically Upregulated in Tissues Related to Totipotency. Mol Biotechnol 48:156–164

Ting NC, Jansen J, Nagappan J, Ishak Z, Chin CW, Tan S-G, Cheah S-C, Singh R (2013) Identification of QTLs associated with callogenesis and embryogenesis in oil palm using genetic linkage maps improved with SSR markers. PLoS ONE 8(1):e53076

Tranbarger TJ, Kluabmongkol W, Sangsrakrul D, Morcillo F, Tregear JW, Tragoonrung S, Billotte N (2012) SSR markers in transcripts of genes linked to post-transcriptional and transcriptional regulatory functions during vegetative and reproductive development of *Elaeis guineensis*. BMC Plant Biol 12:1

Tregear JW, Morcillo F, Richaud F, Berger A, Singh R, Cheah SC, Hartmann C, Rival A, Duval Y (2002) Characterization of a defensing gene expressed in oil palm inflorescences: induction during tissue culture and possible association with epigenetic somaclonal variation events. J Exp Bot 53:1387–1396

Wooi KC (1995) Oil palm tissue culture—current practice and constraints. In: Rao V, Henson IE, Rajanaidu N (eds) Proceedings of the 1993 ISOPB international symposium on recent developments in oil palm tissue culture and biotechnology. Palm Oil Research Institute of Malaysia, Kuala Lumpur, pp 21–32

Zamzuri I (2004) Clonal trial report. Biology Division, MPOB. 26 March 2004

Zamzuri I (2011) MPOB clonal propagation programme. Paper presented at ISOPB international seminar on breeding for sustainability in oil palm. Kuala Lumpur, 18 November 2011

Zamzuri I, Siti Rahmah AR (2007) Tissue culture of *Elaeis oleifera*. Paper presented at PIPOC 2007—poster presentation

Oil Palm Transgenic Research: Challenges, Update, and Future Outlook

6

Omar Abdul Rasid, Subhi Siti Masura, Ayub Nor Hanin, Dayang Izawati Abang Masli, Bahariah Bohari, Mat Yunus Abdul Masani, and Ghulam Kadir Ahmad Parveez

Abstract

Oil palm is a major economic crop for Malaysia and it has contributed significantly to the country's revenues. However, the industry is facing a number of challenges including labor shortage, reduced availability of arable land, and competition from other sources of oils and fats. Thus, strategies must be put in place for the industry to remain competitive. This could be attained through increase in yield, diversification of palm oil applications into higher value products, and production of value-added products. Genetic engineering has been identified as a complementary tool to breeding technique for overcoming the above challenges. The oil palm genetic engineering program was initiated more than two decades ago with the main target to increase oleic acid content. The initial optimization work for oil palm transformation was carried out using biolistics protocol leading to the first successful report of oil palm transformation in the mid 1990s. The optimized protocol was later used to produce transgenic oil palm with high oleic acid as well as other target products. However, analysis of these plants indicated the low efficiency and repeatability of the protocol. Work to improve the transformation efficiency was subsequently carried out including optimization of selection scheme. Moreover, due to recent successful application of *Agrobacterium*-mediated transformation in several monocots, effort to develop the system for oil palm was also initiated. This paper describes the efforts to develop the transformation systems for oil palm and challenges faced during the process.

6.1 Introduction

Oil palm is the main commodity crop of Malaysia. Since it was first introduced as a commercial crop in 1917, the planted area and palm oil production have continuously increased and reached 5.81 million hectares and 19.92 million tonnes in 2017, respectively. The crop has continuously and significantly contributed to the country's economic development and foreign exchange earnings. In 2017, the total export value of palm oil and its products was recorded at RM 77.85 billion, representing 14.6% increase from the previous year (Kushairi 2018). However, Malaysian palm oil industry is facing several challenges. Among the current prominent challenges are the decrease in land availability

O. A. Rasid (✉) · S. S. Masura · A. N. Hanin · D. I. A. Masli · B. Bohari · M. Y. A. Masani · G. K. A. Parveez
Malaysian Palm Oil Board, No. 6, Persiaran Institusi, Bandar Baru Bangi, 43000 Kajang, Selangor, Malaysia
e-mail: omar@mpob.gov.my

© Springer Nature Switzerland AG 2020
M. Ithnin and A. Kushairi (eds.), *The Oil Palm Genome*, Compendium of Plant Genomes, https://doi.org/10.1007/978-3-030-22549-0_6

for further expansion and labor shortage (Khoo and Chandramohan 2002; Cheah 2000; Parveez et al. 2000; Sambanthamurthi et al. 2000). In addition, Malaysian palm oil industry is also faced with competition from other oils such as soybean oil, rapeseed oil, and sunflower seed oil as well as other palm oil producers. The challenge for oil palm industry is to ensure the demand for its oil continues to increase.

In the past, the competitive edge has been the driving force to the rapid growth and tremendous performance of the Malaysian palm oil industry. This will remain as a vital factor for its future development. Therefore, proper strategies should be put in place in order to meet the current challenges and to remain competitive in the future. The industry should increase its productivity, look into the opportunities to diversify its income sources, and expand the end-use base for palm oil. A straightforward and effective approach toward increasing the productivity is to improve the oil yield for each unit of planted area. In the past, improvement has been attained through traditional breeding. The introduction of tenera hybrid as the commercial planting materials has been responsible for the 30% increase of the oil yield compared to dura (Soh et al. 1994). Recently, new tenera varieties with a higher oil yield potential (8–10 t/ha/yr) have been obtained through work in advanced breeding technology. It is believed that further selection of newer elite duras and pisiferas derived from selected germplasm collection can further increase the tenera oil yield in the future (Kushairi et al. 2011; Rajanaidu et al. 2007). Despite the success, the technique has a number of limitations. Improvement of perennial crop such as oil palm that can be achieved through breeding is rather limited. Species incompatibility limits the genetic resource only to very close relatives which can be introgressed into current planting materials. Another serious limitation of this technique is the long generation cycle of the crop. It may take 10–15 years to acquire yield records and to test the consistency of the introgressed characteristics. On top of that, breeding trials usually require a dedication of large land areas (Sambanthamurthi et al. 2000). This is further confounded by the fact that most of the economically important traits are under the control of several genes resulting in difficulty to introgress even single traits (Sambanthamurthi et al. 2009).

Improvement of oil palm planting materials through genetic engineering is believed to be able to enhance the competitiveness and ensure sustainability of the industry (Basiron and Chan 2005; Sambanthamurthi et al. 2000). The feasibility and usefulness of the techniques to improve agronomic traits have been numerously shown in other plants. These include transgenic plants for disease resistance, herbicide resistance, modification of fatty acid contents, or other metabolites and production of new compounds such as certain polymers (Murphy 2014). In addition to speeding up trait improvement, genetic engineering also removes other major impeding factors normally faced by breeders. The source of genes is no longer limited to compatible species. They may come from any organism even in different phyla. This, in theory, could provide large genetic resources to be utilized. The technique also allows more precise alterations to be made on the plant genome compared to conventional breeding. Furthermore, this technology can improve the gene introgression's precision by restricting the amount of genetic materials transferred and ultimately reduce the cost and time for introgressing the desired characteristics. With such advantages, application of genetic engineering for oil palm improvement is pivotal to the industry's future. The application of this technology also makes the penetration and expansion of palm oil into newer markets, such as pharmaceutical and nutraceutical possible (Basiron and Chan 2005; Sambanthamurthi et al. 2000).

The work to genetically oil palm was initiated in the late 80s. With the advancement in genetic modification techniques, the effort to genetically improve oil palm was escalated in the 1990s. In addition, the technology has also been suggested to be used for increasing oil yield, modifying oil composition and developing oil palm with tolerance to pests and diseases (Murphy 2014). These targeted traits are very essential to increase the overall output from the oil palm industry and for venturing into newer markets.

Prior development of several tools and inputs from several fields are necessary for the success of oil palm genetic modification. This includes comprehensive biochemical studies on pathways of interest, gene and promoter isolation for the required genetic materials, construction of transformation vectors, and establishment of foreign DNA delivery system. This article provides an overview of the development of the transformation system for oil palm. It discusses the challenges faced during the development and highlights the recent updates.

6.2 Development of DNA Delivery Procedures

Effort to develop the transformation system for oil palm was initiated in the early 90s focusing on the attempts to establish the gene delivery procedures. Due to low rate of success achieved with *Argobacterium* in monocots then, the development of oil palm transformation system was primarily focused on a direct DNA transfer method, biolistics. This method was specifically developed as a DNA delivery system for plants that were recalcitrant to *Agrobacterium* transformation. The success of DNA delivery is largely independent of target tissue genotype thus allowing any type of cell or tissue to be used. Another obvious advantage of biolistics is that the method allows transfer of large DNA fragments. It has been shown that DNA fragment of up to 150 kb could be transferred into plant cells.

The early works in oil palm transformation were to optimize physical and biological parameters, identify the most effective constitutive promoter, and determine the concentration of selection agents to select transformed oil palm cells (Parveez et al. 1996, 1997, 1998; Chowdhury et al. 1997). The success of oil palm transformation was first reported in mid 1990s following the comprehensive optimization work. Transformed oil palm embryogenic calli (EC) obtained using the optimized condition were later cultured on a selection medium containing the herbicide Basta (glufosinate ammonium as the active ingredient) for the regeneration of putative transgenic plantlets, which were subsequently confirmed by molecular and protein analyses (Parveez 2000). Despite this success, the developed system has a number of limitations. The optimized protocol was shown to have a very low efficiency at about 1.5% based on transient expression. This posed a serious problem either in the optimization work or actual production of transgenic palm for any particular agronomic trait. For any target trait, a large amount of EC needs to be bombarded in order to increase the chance of getting the transgenic plants. Similarly, the optimization of biolistics protocol involved a number of different parameters. Therefore, a large number of EC are also required to be transformed to test all the parameters. This resulted in the extensive tissue culture process which needs a large work force to maintain or subculture the EC, prepare the media for subculture, and handle the GM wastes. In addition to work force, a large space is also required to maintain the cultures, regenerated shoots as well as plantlets for rooting. Unlike most other plants, oil palm has a long regeneration time. Typically, it takes about 18–24 months to regenerate plantlets from EC, another reason for a large work force and space requirement (Fig. 6.1).

Due to this low efficiency, the delivery protocol using biolistics was reevaluated. A number of parameters are currently being assessed or evaluated again toward a higher efficiency based especially on stable transgene integration. Due to the long regeneration time of oil palm tissue cultures, it is expected to take some time to develop DNA delivery system with improved efficiency (Parveez et al. 2017).

Embryogenic calli have been used as the target materials for the optimization and for the succeeding work. EC provides an excellent source of dividing cells which is required for efficient regeneration. However, the regeneration of EC is genotype-dependent. The use of EC with different genetic background could have been one of the reasons for low repeatability observed in oil palm transformation work.

a) b)

Fig. 6.1 **a** Work to develop the protocol to transform oil palm is time and labor intensive. The work requires a big group of supporting staff to subculture and maintains the calli to produce plantlets. **b** A large space is also required to maintain the cultures to regenerate transgenic plantlets. The process usually takes between 18 and 24 months of selection and regeneration

Use of immature embryos (IE) has been widespread in other plants such as soybean, maize, and wheat. IEs are abundant in oil palm and have a great potential to be used as target material. Usually, up to 1000 IEs could be obtained from a single fruit bunch. However, the use of IEs as the target tissue in oil palm transformation work has serious disadvantages. The genotype and the performance of the IEs are distributed according to normal curve. Therefore, the rate of callusing, regeneration, and transformation is not consistent from one IE to another.

Although the biolistics method of DNA delivery into plant cells has been successfully used in a number of plant species, there are a number of drawbacks associated with the method. In general, the method is known to result in insertion of multiple copies of transformed gene(s) into the genome of transgenic plants. In addition, the method is also prone to transgene rearrangement and vector insertion into the genome of transformed plants. These factors usually promote the silencing of gene of interest. *Agrobacterium*-mediated transformation, on the other hand, is known to result in the introduction of single or low copy of the transgene(s). However, transformation of monocots using this method was still very rare in the early 1990s possibly due to host-range of the bacteria (Chan et al. 1993; Ishida et al. 1996). However, as early as 1996, the method started to be applicable in some monocots including a few recalcitrant species such as maize (Ishida et al. 1996) and wheat (Cheng et al. 1997). This was attained after extensive optimization of parameters involved. The feasibility of *Agrobacterium*-mediated transformation in these monocots has led to the initiation of *Agrobacterium*-mediated transformation for oil palm. The first report of successful transformation of oil palm mediated by *Agrobacterium* was reported almost ten years later (Masli et al. 2009). Nevertheless, the efficiency of the developed protocol was rather low at 0.7–1.0% compared to transformation efficiency in other monocots. This result signals that a more comprehensive parameter optimization is needed. As in the case of biolistics, this would require testing a large number of parameters and thus would also require a large work force and space.

The success of *Agrobacterium*-mediated transformation system in monocotyledonous plants is influenced by factors such as *Agrobacterium* strain, binary vector, plant genotype, explant used, selectable marker gene, promoter, inoculation and co-culture conditions, and tissue culture and regeneration medium (Cheng et al. 2004). In oil palm, most of the *Agrobacterium*-mediated transformation works were done on variety tenera (Abdullah et al. 2005; Fuad et al. 2008; Masli et al. 2009; Boonyaves et al. 2009; Ismail et al. 2010; Yenchon and Te-chato 2012; Promchan and Te-chato 2013; Izawati et al. 2015). In 2005, Abdullah et al. demonstrated that immature embryos from all three *Elaeis guineensis* varieties, namely dura, pisifera and tenera as well as from *Elaeis oleifera* were amenable for *Agrobacterium*-mediated transformation system but no comparison were reported for embryogenic calli. In other crops such as wheat, transformation and regeneration of infected explants are highly genotype-dependent. The largest transformation efficiency was reported for breeding line “Bobwhite” compared to other cultivars (Pérez-Piñeiro et al. 2012). In soybean, protocols for transformation were also genotype-dependent (Somers et al. 2003). Later, an improved protocol of soybean transformation by Paz et al. (2004) has made twelve soybean cultivars amenable for genetic improvement.

In the early years, IEs were used as the explant for *Agrobacterium*-mediated oil palm transformation (Abdullah et al. 2005; Fuad et al. 2008; Ismail et al. 2010). Ideally, IEs are the potential explant because they are abundant. Nevertheless, they are often non-uniform in terms of genetic make-up because IEs are products of cross-pollination (Abdullah et al. 2005). Therefore, embryogenic suspension cultures (ECs) were also evaluated as the target tissue (Masli et al. 2009; Boonyaves et al. 2009; Yenchon and Te-chato 2012; Promchan and Te-chato 2013; Izawati et al. 2015). ECs are believed to be the most ideal explants for *Agrobacterium* transformation of oil palm because it is a single cell. This would potentially avoid the occurrence of chimerism in the regenerated plants from genetic transformation (Huang et al. 2007).

In addition to target tissue, other factors that could affect the success of *Agrobacterium*-mediated transformation of oil palm such as *Agrobacterium* strains, bacteria density, inoculation and co-cultivation period, vectors, and selection agent were also studied (Boonyaves et al. 2009; Promchan and Te-chato 2013; Fuad et al. 2008; Masli et al. 2009; Izawati et al. 2015). For *Agrobacterium* strains, LBA4404 is the most widely used strain followed by EHA101, AGL-1, and EHA105. EHA101 was superior to AGL-1 as reported by Promchan and Te-chato (2013). However, there is no report to compare the efficiency of these strains in infecting oil palm explant concurrently. Yenchon and Te-Chato (2012) demonstrated that the highest levels of transient *gus* expression and hygromycin-resistant ECs were obtained using *Agrobacterium* density of 0.8 (OD_{600}) and co-cultivated for 3 days. Thus, the potential of each strain in oil palm transformation as well as other critical factors that could improve the transformation efficiency are currently being evaluated.

In addition to biolistics and *Agrobacterium*-mediated, a transformation system based on protoplast has also been assessed for oil palm. This was possible following the successful regeneration of oil palm from protoplasts which was recently reported (Masani et al. 2013). Some evidence of transgene introduction into protoplasts using PEG and microinjection was later reported (Masani et al. 2014). The study has opened up the possibility of regenerating transgenic oil palm plant from a single cell. In addition, complete elimination of chimera and escapes could also be achieved once the system is fully developed.

6.3 Selection and Regeneration of Transformed Oil Palm Cells

In addition to delivery of foreign DNA into plant cells, selection and regeneration procedure of the cells into a complete plant are the next absolute prerequisites for any genetic modification work. This is usually carried out using a selectable marker that confers resistance to an antibiotic or herbicide. The presence of the selectable marker allows regeneration of transformed cells on medium supplemented with the corresponding antibiotic or herbicide. In oil palm, the initial transformation studies were carried out using *bar* as selectable marker and Basta as the selection agent (Parveez et al. 2007). The gene confers resistance to phosphinotricin (PPT), the active ingredient of the herbicide Basta (DeBlock et al. 1987). The transgenic status of plants regenerated from the initial work was confirmed by molecular and protein analyses. The developed protocol was later used to transform oil palm calli and to regenerate transgenic oil palm plants for a number of genes of interest. Unfortunately, molecular analysis of the regenerated plants revealed the presence of chimeric plants and escapes among these plants (Nurfahisza et al. 2014). The result suggested the low repeatability of the system and could have been resulted from the use of calli with different genotype backgrounds or different callus lines. The presence of escapes and chimeric plants among the regenerated plantlets could also be due to non-optimal selection process. Although the problem is quite common in other plants, it is more prevalent in oil palm (Fig. 6.2).

The finding warranted a reevaluation of the selection scheme used to generate transgenic oil palm. The reassessment of the selection scheme is necessary in order to eliminate or greatly reduce the escapes and chimera, and subsequently improve the efficiency of oil palm transformation system as well. There are a number of factors or parameters that are being looked into including different selectable markers as well as the corresponding selection agents and evaluation of different stages of oil palm tissue culture, with the intention to tighten the selection procedure at every stage of callus development. For the *bar* selectable marker, the effectiveness of two other selection agents, namely glufosinate ammonium and bialaphos has been investigated. Results indicated that these two selection agents were also effective for oil palm. A low concentration of these selection agents is required to inhibit the growth of oil palm calli and embryoids (Nurfahisza et al. 2016). In addition to *bar*, the potential of hygromycin phosphotransferase (*hpt*) selectable marker gene is also currently being evaluated for oil palm transformation. The *hpt* gene was isolated from *Escherichia coli* and confers resistance to the antibiotic hygromycin (Gritz and Davies 1983). Research has been conducted to determine the effective concentrations of hygromycin for different stages of oil palm tissue culture development. The result which was recently reported indicated that the antibiotic was very effective on oil palm calli and therefore has a very high potential to be used for selection and regeneration of oil palm transgenic plants (Fakhrana et al. 2019).

As indicated above, mature or embryogenic calli have been the primary target tissues in the previous oil palm transformation work using biolistics. Recently, the potential of immature calli as the target tissues for oil palm transformation has also been assessed. These are calli at the early stage of tissue culture process and still in callogenesis phase. The friability and lack of

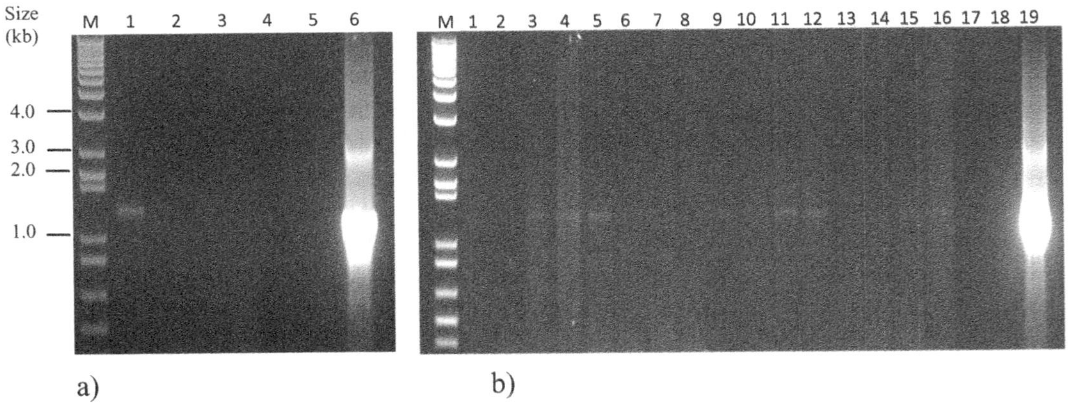

Fig. 6.2 **a** Chimeric nature of some of the regenerated transgenic oil palms. PCR analysis of *bar* gene indicated that leaves from only one (lane 1) of the four fronds tested were positive. Leaves from the other three fronds (lanes 2, 3, and 4) did not carry the transgene. Lanes 5 and 6 are negative and positive controls, respectively. **b** Presence of escapes among the regenerated transgenic oil palm. A large number of the regenerated oil palms were positive for the *bar* gene. However, a few of the samples did not carry the transgene such as samples in lanes 13, 14, and 17. Lanes 18 and 19 are negative and positive controls, respectively. M is 1 kb plus DNA ladder

differentiation of immature calli in early callus proliferation may facilitate an effective *in vitro* selection, thereby minimizing the chance of escapes and chimera. This new strategy has resulted in some positive outcome. Regeneration of oil palm plantlets resistant to glufosinate ammonium from immature EC was successfully obtained. Initial molecular analysis *via* PCR indicated that some of the plantlets carrying the high oleate traits were positive for *bar* gene (Masura et al. 2017). Nevertheless, despite the attempts made to eliminate or reduce the number of escapes or chimera, the problem was still quite persistent as indicated by the above result. Therefore, further optimization of the DNA delivery system and fine tuning of the selection scheme are absolutely necessary and are currently being intensively pursued.

The use of nondestructive visual reporters such as green fluorescent protein gene (*gfp*) to screen and select transformed cells has been successfully demonstrated in a number of plants such as sugarcane (Elliot et al. 1998), oat (Kaeppler et al. 2000), rice (Vain et al. 2000), bromegrass (Nakamura and Ishikawa 2006), and petunia (Mubmann et al. 2011). Attempts to couple this nondestructive visual reporter and selectable marker genes to produce an effective transformation system for oil palm were initiated. In particular, *gfp* gene was evaluated for its suitability to be used in oil palm. A number of factors and parameters associated with marker gene were assessed including different GFP versions, promoters, backbone vectors, plasmid size, and organelle-targeted (Parveez and Majid 2008; Majid and Parveez 2007). However, the efforts have so far been unsuccessful in producing transgenic oil palm expressing the GFP genes in whole tissues (Majid and Parveez 2016).

Selection and regeneration of transformed cells into transgenic plants using selectable marker genes that confer resistance to antibiotics and herbicides are not desirable for commercialization. The presence of these selectable marker genes in transgenic plants could raise concerns over their potential impacts on environment and food safety (Miki and McHugh 2004; Rai and Shekhawat 2014). As such, attempts were later made to use positive selectable markers for the regeneration of transgenic oil palm.

Efforts to use positive selection agents in oil palm were initiated using the *pmi* gene encoding phosphomannose isomerase (PMI) with mannose as the selection agent (Bahariah et al. 2012, 2013). Mannose is converted to mannose-6-phosphate by endogenous hexokinases and cannot be metabolized by plants. PMI however converts mannose-6-phosphate into fructose-6-phosphate (Reed et al. 2001; Bahariah et al. 2012, 2013). The *pmi* gene in transgenic cells thus allows them to survive on medium containing mannose as the carbon source while, on the other hand, untransformed cells will eventually die due to the accumulation of toxic mannose-6-phosphate. Another positive selection system tested in oil palm was 2-deoxyglucose (2-DOG) as a selection medium with the $DOG^{R}1$ gene that encodes 2-deoxyglucose-6-phosphate phosphatase and confers resistance to 2-DOG (Masli et al. 2012; Izawati et al. 2015). 2-DOG prevents cell development when it is converted to 2-DOG-6-phosphate by the phosphorylation of cytosolic hexokinase (Kunze et al. 2001).

6.4 Genes and Promoters

The availability of genetic materials is another imperative requirement for any genetic engineering work. The work to make these genetic materials available was started in the early 1990s. Genes that are related to the pathways of interest such as fatty acid and carotenoid synthesis have been isolated and characterized. For fatty acid synthesis, the genes isolated include acetyl-CoA carboxylase (Wan Saridah et al. 2008), 3-keto-acyl-ACP synthase II (KAS-II) (Umi Salamah and Sambanthamurthi 1996), palmitoyl-ACP thioesterase (Abrizah et al. 2000; Parveez et al. 2010), stearoyl-ACP desaturase (Shah et al. 2000), oleoyl-ACP thioesterase (Asemota et al. 2004), oleoyl-ACP desaturase (Syahanim et al. 2007) and lysophosphatidic acid acyltransferase (Manaf et al. 2005). Meanwhile, isolation of genes involved in carotenoid synthesis including phytoene synthase (Rasid et al. 2008), phytoene desaturase (Rasid et al. 2014), lycopene cyclases (Rasid et al. 2009), and 1-deoxy-D-xylulose-5-phosphate synthase (Khemvong and Suvachittanont 2005) has also been reported. In addition, genes involved in lipid breakdown such as lipase (Nurniwalis et al. 2007, 2015; Morcillo et al. 2013) have also been isolated. Moreover, toward understanding the mechanism of fruit abscission, the key genes such as aminocyclopropane-1-carboxylic acid synthase (ACC synthase) (Tranbarger et al. 2011) and polygalacturonase (Roongsattham et al. 2012) have also been isolated and characterized. The expression and function have also been studied for most of the isolated genes.

In addition to the genes, the works to isolate the regulatory sequences or promoters were also carried out. This includes a number of tissue specific promoters such as mesocarp specific (MT3-A) (Siti Nor Akmar and Zubaidah 2008) and FLL1 (Nurniwalis et al. 2015), kernel specific (pOP-KT21) (Siti Nor Akmar et al. 2014), root specific (MT3-B) (Zubaidah and Siti Nor Akmar 2005; 2010), and leaf-specific (LS01) (Chan et al. 2008; Hanin et al. 2016). The constitutive promoters were isolated from ubiquitin extension protein (*uep*1) (Masura et al. 2010) and translationally control tumor protein (TCTP) genes (Masura et al. 2011). The updates on the works of genes and promoters isolation have been detailed in a number of recent publications (Parveez et al. 2015a; Masura et al. 2017; Mohamad Arif et al. 2017).

6.5 Updates on Generation of Transgenic Oil Palm

The strategy and regeneration of transgenic oil palm for the targeted traits have been detailed in recent review papers (Mohamad Arif et al. 2017; Parveez et al. 2015a). The modification of fatty acid biosynthesis pathway for production of high oleic oil has always been the main priority of oil palm genetic engineering program. The high oleic oil would be the best feedstock for the well expanding oleochemical industry. The less saturated oil would also allay the current fear of consuming saturated fats and the oil will be more liquid, able to be used as a salad oil. Other products targeted are high stearic acid, biodegradable plastic, pharmaceuticals, and nutraceuticals such as lycopene and palmitoleic acid, and industrial oils such as ricinoleic acid (Parveez 2003, 2015b; Sambanthamurthi et al. 2000). The list of target traits, including high oleic acid, the main focus and their potential uses as well as the potential genes involved can be found in a recent review paper (Mohamad Arif et al. 2017).

In general, a number of different transformation vectors were designed and constructed for each of the traits above. Oil palm embryogenic cultures were bombarded with these constructs, which usually carried basta resistance gene. The bombarded embryogenic calli were cultured on selection plates to generate resistant embryogenic calli, which later developed into plantlets. Some of the plantlets have already been transferred to soil and grown in the biosafety nursery. Initial molecular analyses, using PCR and southern hybridization, showed that a few of the plantlets carried the transgenes (Nurfahisza et al. 2014). However, the results also revealed the presence of chimeric transgenic palms. The findings suggest the need for further improvement of the developed transformation protocols.

6.6 The Way Forward and Conclusion

The efforts to genetically engineer oil palm for different oil composition using the outlined strategies are still ongoing. Concurrently, novel approaches are being sought to speed up the achievement of all the objectives. A number of recent development in the area of biotechnology that could be adopted or exploited to facilitate efforts to genetically engineer oil palm are discussed below.

Recently, the oil palm genome and other omics data have been reported (Singh et al. 2013; Ramli et al. 2017; Nurazah et al. 2017). Their potential applications in oil palm biotechnology have been recently reviewed (Masura et al. 2017; Parveez et al. 2017). The availability of these data is extremely useful to oil palm genetic engineering work. The data can be utilized in many ways which could speed up the progress of the program. In particular, the data can be used to identify more genes and promoters, or any other genetic elements required in the genetic engineering work. For example, a number of genes that are preferentially expressed in mesocarp tissues were identified by analyzing oil palm transcriptomics data (Siti Suriawati et al. 2016, 2017). This would be extremely useful for the isolation of mesocarp or any other tissue-specific promoters from oil palm. The available data would also allow the identification of members of a gene family. This would simplify the procedure to genetically engineer a trait that is controlled by a gene family. A specific member of the given gene can be identified and targeted for a precise result.

Another recent development that could potentially be adopted in oil palm genetic engineering work is the genome editing. This is a technology that can potentially be applied in any type of cells and organisms (Gaj et al. 2013). In general, the technology targets specific DNA sequences by introduction of customized nucleases carrying sequence-specific DNA-binding domains (Wyman and Kanaar 2006). The two well established and commonly used DNA domain nucleases for genome editing approach are zinc-finger nucleases (ZFN) and transcription activator-like effector-based nucleases (TALENs). However, these technologies suffer from the complicated vector design, particularly in assembling the relevant DNA-binding protein for each gene of interest (Belhaj et al. 2013). In addition, the time-consuming factor makes these approaches not widely adopted for plant gene regulation. Recently, a simpler method based on the bacterial type II Clustered Regularly Interspaced Short Palindromic Repeats (CRISPR)/Cas (CRISPR-associated) immune system was developed. This method allows cleavage of genomic DNA by nucleases at a targeted site which is guided by a customized small noncoding RNA. A number of reviews have highlighted the applications of this technique in crop plants (Jung et al. 2018; Malzhan et al. 2017; Zhang et al. 2017). The main advantage of this technology is the less regulatory scrutiny on the developed products. Some regulatory authorities, such as USDA, have decided that plants mutated by genome editing, should not be regulated (Wolt et al. 2016). Moreover, mutations created through genome editing are stable and heritable over generations. The availability of oil palm genome and transcriptome data could facilitate the development and application of this technique for trait improvement.

Another recent breakthrough that can potentially be applied in oil palm is the use of growth stimulating genes to improve the recovery of transformed plants. Overexpression of these genes, namely *Baby boom* (*Bbm*) and *Wuschel2* (*Wus2*), was shown to increase the transformation efficiency of recalcitrant maize inbred lines. The feasibility of this breakthrough was also extended to other monocots, namely sorghum, sugarcane and indica rice (Lowe et al. 2016). This technology has also been demonstrated to allow the regeneration of transgenic plants without utilizing a selectable marker gene (Mookkan et al. 2017). This latest finding could potentially increase the acceptance of genetically modified products in general.

Acknowledgements The authors wish to thank the MPOB management for permission to publish this chapter.

References

Abdullah R, Zainal A, Wee YH, Leaw CL, Yeap CB, Lee MP, Sirajudin SA, Yap SPW, Joseph JL, Jusoh SA, Muad MR, Yeun LH (2005) Immature Embryo: a useful tool for oil palm (*Elaeis guineensis* Jacq.) genetic transformation studies. Electron J Biotechnol 8(1):25–34

Abrizah O, Lazarus C, Fraser T, Stobart K (2000) Cloning of palmitoyl-ACP thioesterase from oil palm. Biochem Soc Trans 28:619–622

Asemota O, San CT, Shah FH (2004) Isolation of a kernel oleoyl-ACP thioesterase gene from the oil palm *Elaeis guineensis* Jacq. African J Biotechnol 3(3):199–201

Bahariah B, Parveez GKA, Khalid N (2012) Determination of optimal concentration of mannose as a selection agent for selecting transformed oil palm cells using the *phosphomannose isomerase (pmi)* gene as the positive selectable marker. J Oil Palm Res 24:1250–1259

Bahariah B, Parveez GKA, Masani MYA, Masura SS, Khalid N, Othman RY (2013) Biolistic transformation of oil palm using the *phosphomannose isomerase* (pmi) gene as a positive selectable marker. Biocatal Agri Biotech 2:295–304

Basiron Y, Chan KW (2005) The role of research and development strategies in food safety and good agricultural, manufacturing and distribution practices in the Malaysian palm oil industry. Oil Palm Indus Econ J 5:1–16

Belhaj K, Chaparro-Garcia A, Kamoun S, Nekrasov V (2013) Plant genome editing made easy: targeted mutagenesis in model and crop plants using the CRISPR/Cas system. Plant Methods 9:39

Boonyaves K, Romyanon K, Kirdmanee C, Supaibulwatana K (2009). *Agrobacterium* tumefaciens-mediated genetic transformation in oil palm (*Elaeis guineensis* Jacq.). Agricultural biotechnology international conference: agricultural biotechnology for better living and a clean environment, Queen Sirikit National Convention Center, Bangkok, Thailand, 22–25 September 2009

Chan MT, Chang HH, Ho SL, Tnag WF, Yu SM (1993) *Agrobacterium*-mediated production of transgenic rice plants expressing a chimeric a-amylase promoter/β-glucuronidase gene. Plant Mol Biol 22:491–506

Chan PL, Siti Nor Akmar A, Roohaida O (2008) Light-harvesting chlorophyll A/B binding protein (LHCB) promoter for targeting specific expression in oil palm leaves. J Oil Palm Res (Spec Iss July), pp 21–29

Cheah SC (2000) Biotechnological strategies for improving plantation tree crop: the palm oil—a case study. In: Proceeding international planters conference, pp 59–76

Cheng M, Fry JE, Pang S, Zhou H, Hironaka CM, Duncan DR, Conner TW, Wan Y (1997) Genetic Transformation of wheat mediated by *Agrobacterium tumefaciens*. Plant Physiol 115(3):971–980

Cheng M, Lowe BA, Spencer TM, Ye X, Armstrong CL (2004) Factors influencing *Agrobacterium*-mediated transformation of monocotyledonous species. In Vitro Cell Dev Biol-Plant 40:31–45

Chowdhury MKU, Parveez GKA, Saleh NM (1997). Evaluation of five promoters for use in transformation of oil palm (*Elaeis guineensis* Jacq.) Plant Cell Rep 16:277–281

DeBlock M, Botterman J, Vanderwiele M, Montagu M, Leeman J (1987) Engineering herbicide resistant in plant by detoxyfying enzyme. EMBO J 6:2513–2518

Elliot AR, Cambell JA, Brettell RIS, Grof CPL (1998) *Agrobacterium*-mediated transformation of sugarcane using GFP as a screenable marker. Aust J Plant Physiol 25:739–743

Fakhrana I, Nurfahisza AR, Rasid OA, Parveez GKA (2019) Minimal inhibitory concentration of hygromycin for selecting transformed oil palm embryogenic calli. J Oil Palm Res 31:14–27

Fuad FAA, Ismail I, Sidik NM, Zain CRCM, Abdullah R (2008) Super binary vector system enhanced transformation frequency and expression level of polyhydroxyvalerate gene in oil palm immature embryo. Asian J Plant Sci 7(6):526–535

Gaj T, Gersbach CA, Barbas CF (2013) ZFN, TALEN, and CRISPR/Cas-based methods for genome engineering. Trends Biotechnol 31:397–407

Gritz L, Davies J (1983) Plasmids encodes hygromycin B resistance: the sequence of hygromycin B phosphotransferase gene and its expression in *Escherichia coli* and *Saccharomyces cerevisae*. Gene 25:179–188

Hanin AN, Masani MYA, Parveez GKA (2016) Evaluation of oil palm leaf-specific promoter (LSP1) activity for expressing PHB genes in *Arabidopsis thaliana*. J Oil Palm Res 28(1):1–9

Huang X, Huang XL, Xiao W, Zhao JT, Dai XM, Chen YF, Li XJ (2007) Highly efficient *Agrobacterium*-mediated transformation of embryogenic cell suspensions of *Musa acuminate* cv. Mas (AA) via a liquid co-cultivation system. Plant Cell Rep 26:1755–1762

Ishida Y, Saito H, Hiei Y, Komari T, Kumashiro T (1996) High efficiency transformation of maize (*Zea mays* L.) mediated by *Agrobacterium tumefaciens*. Nat Biotechnol 14:745–750. https://doi.org/10.1038/nbt0696-745

Ismail I, Iskandar NF, Gor MC, Abdullah R (2010) Genetic transformation and molecular analysis of polyhydroxybutyrate biosynthetic gene expression in oil palm (*Elaeis guineensis* Jacq. var Tenera) tissues. Plant Omics J 3(1):18–27

Izawati AMD, Masani MYA, Ismanizan I, Parveez GKA (2015) Evaluation on the effectiveness of 2-deoxyglucose-6-phosphate phosphatase (DOG^R1) gene as a selectable marker for oil palm (*Elaeis guineensis* Jacq.) embryogenic calli transformation mediated by *Agrobacterium tumefaciens*. Front Plant Sci 6:727. https://doi.org/10.3389/fpls.2015.00727

Jung C, Capistrano-Gossmann G, Braatz J, Niharika Sashidhar N, Melzer S (2018) Recent developments in genome editing and applications. Plant Breeding 137:1–9

Kaeppler HF, Menon GK, Skadsen RW, Nuutila AM, Carlson AR (2000) Transgenic oat plants via visual selection of cell expressing green fluorescent protein. Plant Cell Rep 19:661–666

Khemvong S, Suvachittanont W (2005) Molecular cloning and expression of a cDNA encoding 1-deoxy-D-xylulose 5-phosphate synthase from oil palm *Elaeis guineensis* Jacq. Plant Sci 169:571–578

Khoo KM, Chandramohan D (2002) Malaysian palm oil industry at crossroads and its future direction. Oil Palm Indus Econ J 2(2):10–15

Kunze I, Ebneth M, Heim U, Geiger M, Sonnewald U, Herbers K (2001) 2-Deoxyglucose resistance: a novel selection marker for plant transformation. Mol Breed 7:221–227

Kushairi AD (2018) Overview of the Malaysian palm oil industry 2017. Malaysian Palm Oil Board. Available from: http://bepi.mpob.gov.my/images/overview/Overview_of_Industry_2017.pdf. Accessed 31 May 2018

Kushairi AD, Mohd Din A, Rajanaidu N (2011) Oil palm breeding and seed production. In: Mohd Basri W, Choo YM, Chan KW (eds) Further advances in oil palm research (2000–2010), vol 1. MPOB, Bangi, pp 47–93

Lowe K, Wu E, Wang N, Hoerster G, Hastings C, Cho MJ, Scelonge C, Lenderts B, Chamberlin M, Cushatt J, Wang L, Ryan L, Khan T, Chow-Yiu J, Hua W, Yu M, Banh J, Bao Z, Brink K, Igo E, Rudrappa B, Shamseer PM, Bruce W, Newman L, Shen B, Zheng P, Bidney D, Falco SC, Registerlll JC, Zhao ZY, Xu D, Jones TJ, Gordon-Kamm WJ (2016) Morphogenic regulators *Baby boom* and *Wuschel* improve monocot transformation. Plant Cell 28:1998–2015

Majid NA, Parveez GKA (2007) Evaluation of green fluorescence protein (GFP) as a selectable marker for oil palm transformation via transient expression. Asia Pac J Mol Bio Biotech 15:1–8

Majid NA, Parveez GKA (2016) Regeneration of transgenic oil palm carrying *gfp* gene used as a visual selectable marker. J Oil Palm Res 28(4):415–430

Malzhan A, Lowder L, Qi Y (2017) Plant genome editing with TALEN and CRISPR. Cell Biosci 7:21

Manaf MAA, Abrizah O, Umi Salamah R (2005) Characterization of genes encoding key enzymes in oil synthesis in the oil palm. In: Proceeding of the PIPOC 2005 international palm oil congress (agriculture, biotechnology & sustainability), pp 583–606

Masani MYA, Noll G, Parveez GKA, Sambanthamurthi R, Prüfer D (2013) Regeneration of viable oil palm plants from protoplasts by optimizing media components, growth regulators and cultivation procedures. Plant Sci 210:118–127

Masani MYA, Noll G, Parveez GKA, Sambanthamurthi R, Prüfer D (2014) Efficient transformation of oil palm protoplasts by PEG-mediated transfection and DNA microinjection. PLoS ONE 9(5):e96831. https://doi.org/10.1371/journal.pone.0096831

Masli DIA, Kadir APG, Yunus AMM (2009) Transformation of oil palm using *Agrobacterium tumefaciens*. J Oil Palm Res 21:643–652

Masli DIA, Parveez GKA, Ismail I (2012) Optimisation of 2-deoxyglucose concentration for identifying the sensitivity level for oil palm embryogenic calli. J Oil Palm Res 24:1296–1302

Masura SS, Parveez GKA, Ismail I (2010) Isolation and characterization of oil palm constitutive promoter derived from ubiquitin extension protein (*uep1*) gene. New Biotechnol 27:289–299

Masura SS, Parveez GKA, Eng Ti LL (2011) Isolation and characterization of oil palm constitutive promoter derived from translationally control tumor protein (TCTP*)* gene. Plant Physiol Biochem 49:701–708

Masura SS, Tahir NI, Rasid OA, Ramli US, Abrizah O, Masani AMY, Parveez GKA, Kushairi A (2017) Post-genomic technologies for the advancement of oil palm research. J Oil Palm Res 29(4):469–486

Miki B, McHugh S (2004) Selectable marker genes in transgenic plants: applications, alternatives and biosafety. J Biotechnology 107:193–232

Mohamad Arif AM, Masli DIA, Ramli Z, Yunus AMM, Mohammed S, Hwa LF, Wahab NA, Rasid OA, Sambanthamurthi R, Parveez GKA (2017) Biotechnology for diversification and improved resilience of the oil palm. Planter 93(1093):237–249

Mookkan M, Nelson-Vasilchik K, Hague J, Zhang ZJ, Kausch AP (2017) Selectable marker independent transformation of recalcitrant maize inbred B73 and sorghum P898012 mediated by morphogenic regulators BABY BOOM and WUSCHEL2. Plant Cell Rep 36:1477–1491

Morcillo F, Cros D, Billotte N, Ngando-Ebongue G-F, Domonhédo H, Pizot M, Cuéllar T, Espéout S, Dhouib R, Bourgis F, Claverol S, Tranbarger TJ, Nouy B, Arondel (2013) Improving palm oil quality through identification and mapping of the lipase gene causing oil deterioration. Nat Commun 4:2160. https://doi.org/10.1038/ncomms3160

Mubmann V, Serek M, Winkelmann T (2011) Selection of transgenic Petunia plants using the green fluorescent protein (GFP). Plant Cell Tissue Organ Culture 107:483–492. https://doi.org/10.1007/s11240-011-9998-3

Murphy DJ (2014) The future of oil palm as a major global crop: opportunities and challenges. J Oil Palm Res 26:1–24

Nakamura T, Ishikawa M (2006) Transformation of suspension cultures of bromegrass (*Bromus inermis*) by *Agrobacterium tumefaciens*. Plant Cell Tissue Organ Cult 84:293–299

Nurazah Z, Idris AS, Mohd Din A, Mohamad Arif AM, Abrizah O, Ramli US (2017). A metabolic fingerprinting approach to assess metabolome changes in response to oil palm basal stem rot. In: Proceedings of the PIPOC 2017 international palm oil congress (agriculture, biotechnology & sustainability), pp 341–344

Nurfahisza AR, Rafiqah MA, Masani MYA, Hanin AN, Rasid OA, Parveez GKA, Ismail I (2014) Molecular analysis of transgenic oil palm to detect the presence of transgenes. J Oil Palm Res 26:73–80

Nurfahisza AR, Rafiqah MA, Parveez GKA, Rasid OA (2016) Comparison of the effectiveness of basta, bialaphos and glufosinate ammonium for selecting transformed oil palm tissues. J Oil Palm Res 28:247–255

Nurniwalis AW, Siti Nor Akmar A, Chan PL, Mohamad Arif MA (2007). Isolation of a cDNA encoding lipase class 3 family protein from oil palm (*Elaeis guineensis*). In: PIPOC agriculture, biotechnology and sustainability conference, Kuala Lumpur Convention Centre (KLCC), Kuala Lumpur, August 26–30, pp 1011–1020

Nurniwalis AW, Zubaidah R, Siti Nor Akmar A, Zulkifli H, Mohamad Arif MA, Massawe FJ, Chan KL, Parveez GKA (2015) Genomic structure and characterization of a lipase class 3 gene and promoter from oil palm. Biol Plant 59:227. https://doi.org/10.1007/s10535-015-0500-7

Parveez GKA (2000) Production of transgenic oil palm (*Elaeis guineensis* Jacq.) using biolistic techniques. In: Jain SM, Minocha SC (eds) Molecular biology of woody plants, vol 2. Kluwer Academic Publishers, Netherlands, pp 327–350

Parveez GKA (2003) Novel products from transgenic oil palm. Ag Biotech Net 5:1–8 (ABN 113)

Parveez GKA, Majid NA (2008) Factors affecting green fluorescence protein (GFP) gene expression in oil palm after microprojectile bombardment mediated transformation. J Oil Palm Res 20:495–507

Parveez GKA, Chowdhury MKU, Saleh NM (1996) Determination of minimal inhibitory concentration of selection agents for oil palm (*Elaeis guineensis* Jacq.) transformation. Asia Pac J Mol Biol Biotechnol 4:219–228

Parveez GKA, Chowdhury MKU, Saleh NM (1997) Physical parameters affecting transient GUS gene expression in oil palm using the biolistics device. Ind Crops Prod 6:41–50

Parveez GKA, Chowdhury MKU, Saleh NM (1998) Biological parameters affecting transient GUS gene expression in oil palm embryogenic calli via microprojectile bombardment. Ind Crops Prod 8:17–27

Parveez GKA, Masnita MM, Alizah Z, Majid NA, Abdul Masani MY, Haliza HF, Rasid OA, Cheah SC (2000) Transgenic oil palm: production and projection. Biochem Soc Trans 28:969–972

Parveez GKA, Majid NA, Alizah Z, Rasid OA (2007) Determination of minimal inhibitory concentration of selection agents for selecting transformed immature embryos of oil palm. Asia Pac J Mol Biol Biotechnol 15:133–146

Parveez GKA, Abrizah O, Nurhafizah R, Bahariah B (2010) Functional analysis of oil palm palmitoyl-acyl-ACP thioesterase (FatB) gene via down-regulation in a model plant: *Arabidopsis thaliana*. J Oil Palm Res 22:803–813

Parveez GKA, Rasid OA, Masani MYA, Sambanthamurthi R (2015a) Biotechnology of oil palm: strategies towards manipulation of lipid content and composition. Plant Cell Rep 34(4):533–543

Parveez GKA, Bahariah B, Ayub NH, Masani MYA, Rasid OA, Tarmizi AH, Ishak Z (2015b) Production of polyhydroxybutyrate in oil palm (*Elaeis guineensis* Jacq.) mediated by microprojectile bombardment of PHB biosynthesis genes into embryogenic calli. Front Plant Sci 6:598. https://doi.org/10.3389/fpls.2015.00598

Parveez GKA, Singh R, Ong-Abdullah M, Ramli US, Low ETL, Rasid OA, Manaf Mohamad Arif AM (2017) Impact of biotechnology on the sustainable development of the oil palm industry—from research to application. In: Proceedings of the PIPOC 2017 international palm oil congress (agriculture, biotechnology & sustainability), pp 77–86

Paz MM, Shou H, Guo Z, Zhang Z, Banerjee AK, Wang K (2004) Assessment of conditions affecting *Agrobacterium*-mediated soybean transformation using the cotyledonary node explant. Euphytica 136:167–179

Pérez-Piñeiro P, Gago J, Landín M, Gallego PP (2012) *Agrobacterium*-mediated transformation of wheat: general overview and new approaches to model and identify the key factors involved. In: Çiftçi YO (ed) Transgenic plants—advances and limitations, InTech, Croatia. https://doi.org/10.5772/35232. Available from: https://www.intechopen.com/books/transgenic-plants-advances-and-limitations/agrobacterium-mediated-transformation-of

Promchan T, Te-Chato S (2013) Strains of *Agrobacterium* affecting gene transformation through embryogenic cell suspension of hybrid tenera oil palm. J Agric Technol 9(3):669–679

Rai MK, Shekhawat NS (2014) Recent advances in genetic engineering for improvement of fruit crops. Plant Cell Tiss Organ Cult 116:1–15

Rajanaidu N, Kushairi AD, Chan KW, Mohd Din A (2007) Current status of oil planting material in the world and future challenges. In: Proceedings 2007 PIPOC international palm oil congress, pp 503–520

Ramli US, Abrizah O, Singh R, Ong Abdullah M, Sambanthamurthi R, Kushairi AD, Parveez GKA (2017) An integrative proteomics and metabolomics approach to oil palm research. In" Proceedings of the PIPOC 2017 international palm oil congress (agriculture, biotechnology & sustainability), pp 348–350

Rasid OA, Wan Nur Syuhada WS, Nor Hanin A, Masura SS, Zulqarnain M, Ho CL, Sambanthamurthi R, Suhaimi N (2008) RT-PCR amplification and cloning of partial DNA sequence coding for oil palm (*Elaeis oleifera*) phytoene synthase gene. AsPac J Mol Biol Biotechnol 16:17–24

Rasid OA, Parveez GKA, Ho CL, Suhaimi N, Sambanthamurthi R (2009) Plant carotenoids: molecular genetics and regulation. J Oil Palm Res 21:588–601

Rasid OA, Wan Nur Syuhada WS, Nor Hanin A, Singh R, Ho CL, Parveez GKA (2014) Molecular cloning and regulation of oil palm (*E. guineensis* Jacq.) phyteone desaturase in developing mesocarp tissues. J Oil Palm Res 26:37–46

Reed J, Privalle L, Powell ML, Meghji M, Dawson J, Dunder E, Sutthe J, Wenck A, Launis K, Kramer C, Chang YF, Hansen G, Wright M (2001) Phosphomannose isomerase: An efficient selectable marker for plant transformation. Vitro Cell Dev Biol Plant 37:127–132

Roongsattham P, Morcillo F, Jantasuriyarat C, Pizot M, Moussu S, Jayaweera D, Collin M, Gonzalez-Carranza ZH, Amblard P, Tregear JW, Tragoonrung S, Verdeil JL, Tranbarger TJ (2012) Temporal and spatial expression of polygalacturonase gene family members reveals divergent regulation during fleshy fruit ripening and abscission in the monocot species oil palm. BMC Plant Biol 12:150. https://doi.org/10.1186/1471-2229-12-150

Sambanthamurthi R, Parveez GKA, Cheah SC (2000). Genetic engineering of oil palm. In: Yusof B, Jalani BS, Chan KW (eds) Advances in oil palm research. Malaysian Palm Oil Board

Sambanthamurthi R, Rajinder S, Parveez GKA, Meilina OA, Kushairi A (2009) Opportunities for the oil palm via breeding and biotechnology. In: Jain SM, Priyadarshan PM (eds) Breeding plantation tree crops: tropical species. Springer Science & Business Media, New York, pp 377–421

Shah FH, Rashid O, Cha TS (2000) Temporal regulation of two isoforms of cDNAs clones encoding delta 9-stearoyl-ACP desaturase from oil palm (*Elaeis guineensis*). Plant Sci 27:27–33

Singh R, Meilina OA, Low ETL, Mohamad Arif AM, Rozana R, Rajanaidu N, Ooi CLL, Ooi SE, Chan KL, Mohd Amin H, Norazah A, Jayanthi N, Bacher B, Lakey N, Smith SW, He D, Hogan M, Budiman MA, Lee EK, Desalle R, Kudrna D, Goicoechea JL, Wing RA, Wilson RK, Fulton RS, Ordway JM, Martienssen RA, Sambanthamurthi R (2013) Oil palm genome sequence reveals divergence of interfertile species in Old and New worlds. Nature 500 (7462):335–339

Siti Nor Akmar A, Zubaidah R (2008) Mesocarp-specific metallothionein-like gene promoter for genetic engineering of oil palm. J Oil Palm Res (Spec Iss July):1–8

Siti Nor Akmar A, Cheah SC, Nurniwalis AW (2014) Expression regulatory elements. Patent no: US8791330

Siti Suriawati B, Chan PL, Chan KL, Parveez GKA, Rasid OA (2016) Meta analysis of multiple transcriptome data sets reveals mesocarp-specific genes in oil palm. In: 3rd international plant breeding conference, p 79

Siti Suriawati B, Chan PL, Chan KL, Parveez GKA, Rasid OA (2017) Transcriptomic analysis reveals preferentially expressed genes in the mesocarp tissues of oil palm. In: Proceedings of the PIPOC 20017 international palm oil congress (agriculture, biotechnology & sustainability), pp 174–177

Soh AC, Tan H, Ooi LH, Rajanaidu N, Cheah SC, Low FC (1994). Genetic improvement of plantation crops in Malaysia. In: Proceeding 1st national congress on genetics, pp 55–69

Somers DA, Samac DA, Olhoft PM (2003) Recent advances in legume transformation. Plant Physiol 131:892–899

Syahanim S, Abrizah O, Siti Nor Akmar A, Mohamad Arif AM, Ho CL (2007) Cloning of an oleoyl-Coa desaturase from oil palm. In: Proceedings of the PIPOC 2007 international palm oil congress (agriculture, biotechnology & sustainability), pp 1001–1009

Tranbarger TJ, Dussert S, Joët T, Argout X, Summo M, Champion A, Cros D, Omore A, Nouy B, Morcillo F (2011) Regulatory mecha-nisms underlying oil palm fruit mesocarp maturation, ripening, and functional specialization in lipid and carotenoid metabolism. Plant Physiol 156:564–584

Umi Salamah R, Sambanthamurthi R (1996) β–Ketoacyl-ACP synthase II in the oil palm *(Elaeis guineensis* Jacq.) mesocarp. In: Williams JP, Khan UM and Lem NW (eds) Physiology, biochemistry and molecular biology of plant lipids. Kluwer Academic Publishers, Toronto, pp 69–71

Vain P, Worland B, Kohli A, Snape JW, Christou P (2000) Green fluorescent protein (GFP) as a vital screenable marker in rice transformation. Theor Appl Genet 9:164–169

Wan Saridah WA, Laura BW, Chokyun RHA, Anthony JS, Umi Salamah R, Abdul Masani MY, Ghulam KAP, Ravigadevi S (2008) Acetyl-CoA carboxylase from oil palm (*Elaeis guineensis*) mesocarp. J Oil Palm Res 2:97–107

Wolt JD, Wang K, Yang B (2016) The regulatory status of genome-edited crops. Plant Biotechnol J 14:510–518

Wyman C, Kanaar R (2006) DNA double-strand break repair: all's well that ends well. Annu Rev Genet 40:363–383

Yenchon S, Te-chato S (2012) Effect of bacteria density, inoculation and co-cultivation period on *Agrobacterium*-mediated transformation of oil palm embryogenic callus. J Agric Technol 8(4):1485–1496

Zhang K, Raboanatahiry N, Zhu B, Li M (2017) Progress in genome editing technology and its application in plants. Front Plant Sci 8:177. https://doi.org/10.3389/fpls.2017.00177

Zubaidah R, Siti Nor Akmar A (2005) The effects of metal ions on root-specific expression of the oil palm *MT3-B* gene promoter. In: Proceedings of the PIPOC 2005 International palm Oil congress, pp 1104–1110

Zubaidah R, Siti Nor Akmar A (2010) Functional characterization of the oil palm type 3 metallothionein-like gene (MT3-B) promoter. Plant Mol Biol Rep 28(3):531–541. ISSN 0735-9640

Oil Palm Genome: Strategies and Applications

7

Eng-Ti Leslie Low, Norazah Azizi, Mohd Amin Ab Halim, Nik Shazana Nik Mohd Sanusi, Kuang-Lim Chan, Nadzirah Amiruddin, Nagappan Jayanthi, Meilina Ong-Abdullah, Rajinder Singh, Ravigadevi Sambanthamurthi, Mohamad Arif Abdul Manaf, and Ahmad Kushairi

Abstract

Oil palm produces two types of edible oil, namely palm oil and palm kernel oil that are important sources of food and raw materials for industries, such as cosmetics, biofuels, pharmaceutical and oleochemicals. It is harvested all year round and has the highest oil yield per hectare compared to other edible oil crops. Nevertheless, its yields have stagnated for the past three decades and continuous improvements are required to keep up with future demands of the growing world population. The need to improve productivity motivated the research community and the oil palm industry to jointly embrace the latest biotechniques to study and understand the crop. This partnership paved the way for the development of more comprehensive research programmes that resulted in the release of the assembled oil palm genome sequence, identification of genes/markers associated with traits of interest, and development of new tools. The most important successes to date are the identification of two genes linked to important agronomic traits, i.e. shell phenotype and fruit colour, and a gene that causes the abnormal mantled somaclonal variant. These discoveries have since been translated into diagnostic assays that can be used to screen and select planting materials as early as at the seed or nursery stage, allowing planters to only plant palms with the desired traits. This chapter explores the developments leading to the sequencing of the oil palm genome and its subsequent application in the development of new knowledge and tools that will help shape the next generation of planting materials.

7.1 Introduction

Oil palm consists of two important species, namely *Elaeis guineensis* which originates from Africa and *E. oleifera* from South America. It is the most productive oil crop in the world. In 2017, palm oil and palm kernel oil accounted for 38.7% of the world's production of vegetable oils, which totalled ~192.8 million tonnes while using only ~7% of the world's harvested area for oil seeds (MPOB 2017). The productivity of the crop is in part contributed by the discovery of the monogenetic inheritance of the *SHELL* gene (Sh) by Beirnaert and Vanderweyen (1941) in Congo. The discovery changed the face of the oil palm industry and gave a tremendous boost in oil yields. The heterozygous *tenera* (thin shell),

E.-T. L. Low (✉) · N. Azizi · M. A. A. Halim · N. S. N. M. Sanusi · K.-L. Chan · N. Amiruddin · N. Jayanthi · M. Ong-Abdullah · R. Singh · R. Sambanthamurthi · M. A. A. Manaf · A. Kushairi
Malaysian Palm Oil Board, No. 6, Persiaran Institusi, Bandar Baru Bangi, 43000 Kajang, Selangor, Malaysia
e-mail: lowengti@mpob.gov.my

© Springer Nature Switzerland AG 2020
M. Ithnin and A. Kushairi (eds.), *The Oil Palm Genome*, Compendium of Plant Genomes, https://doi.org/10.1007/978-3-030-22549-0_7

which is a hybrid of *dura* (thick shell) and *pisifera* (shell-less), has a higher proportion of oil-bearing pulp or mesocarp compared to *dura* (Kushairi and Rajanaidu 2000). *Pisifera* is normally female sterile and generally does not produce fruits. *Tenera* produces ~30 and 100% more oil than *dura* and *pisifera*, respectively, making *tenera* the most commonly used commercial material in South-East Asia. Nevertheless, further substantial improvements to yield have since been limited and yields have stagnated in the last three decades.

This brought about interest in vegetative propagation of superior and uniform oil palm planting materials (Fig. 7.1) via tissue culture (Wong et al. 1997). Clonal palms have been shown to produce up to 30% more yield than standard crosses of *dura* × *pisifera* materials (Khaw and Ng 1998; Cochard et al. 1999). However, the drive to mass-propagate clonal palms was dampened by two factors—the low rate of embryogenesis in leaf-derived callus that ranged from 3 to 6% (Wooi 1995) and the occurrence of somaclonal variants. Although the cause of a major somaclonal variant phenomenon called mantled (Ong-Abdullah et al. 2015) has been identified, the underlying cause of a number of other somaclonal variants, such as androgynous flowers and sterility, is still unknown. Nevertheless, the identification of the *MANTLED* gene has built confidence in the use of tissue culture to produce new planting materials, as a diagnostic assay is now available to identify abnormal mantled clones in the nursery. Moving forward, there is still a lot that needs to be done to improve efficiency of the tissue culture process.

With this in mind and the need to improve yield, researchers turn to the latest developments in molecular biology, molecular markers and sequencing technologies to unravel the regulatory processes that control important traits in oil palm. Used together with conventional breeding, these technologies provide the framework for the

Fig. 7.1 Uniform growth of clonal palms at United Plantations Berhad's Jendarata Estate, Teluk Intan, Perak, Malaysia

development of tools to implement marker-assisted selection to reduce the long breeding cycle of oil palm, which can take up to 10 years, and plant more precisely selected elite materials. This chapter discusses the developments and applications that led to the sequencing of the oil palm genome and its utilization to improve the efficiency of oil palm breeding.

7.2 Early Application of Sequencing Technology in Oil Palm

The rising popularity of molecular biology and the advent of molecular marker technologies in the 1980s and 1990s provided renewed hope to the oil palm industry. These new sciences were deployed to help identify key genes and markers associated with traits of interest. The initial foray into this was based on targeted gene approaches. This can be seen by research carried out by Adam et al. (2006) and Alwee et al. (2006), where they used publicly available MADS box gene sequences to design primers to identify a set of 13 and 14 oil palm MADS box genes, respectively. These genes proved to be useful in expanding the understanding of the underlying mechanisms in flowering, as further analysis showed that these genes control floral organ identity using the generic dicot ABCDE model (Adam et al. 2007). For tissue culture, Morcillo et al. (2001) were able to correlate the accumulation of 7S globulin with expression of an oil palm 7S globulin protein gene (GLO7A) in oil palm zygotic and somatic embryos. The gene has abscisic acid (ABA)-responsive elements and one motif resembling a seed-specific promoter element in its promoter region, explaining why in vitro production of 7S globulins increases with the addition of arginine, sucrose and ABA (Morcillo et al. 2001). The expression of the gene in somatic embryos was also supported by research carried out by Aberlenc-Bertossi et al. (2008). Other genes identified include MYB-related genes (Teoh et al. 2003) and an AP2/EREBP domain gene (Morcillo et al. 2007).

Although these approaches provided useful information, the throughputs of these methods were limited. The discovery of the expressed sequenced tag (EST) approach by Adams et al. (1991) provided an important tool for gene discovery and expression studies. The breakthrough showed that partially sequenced randomly selected cDNA clones of 150–400 nucleotides were sufficient to tag and identify genes. The EST approach resulted in the discovery and deposition of ~41,200 *E. guineensis* and ~3200 *E. oleifera* sequences in the EST database (dbEST) in GenBank (NCBI 2018). The ESTs were developed from a wide variety of tissues and development stages, such as during micropropagation of oil palm ramets (Ho et al. 2007; Low et al. 2008; Lin et al. 2009; Chan et al. 2010), flower or fruit development (Jouannic et al. 2005; Ho et al. 2007; Nurniwalis et al. 2008; Beulé et al. 2011; Xu et al. 2011), zygotic embryos (Jouannic et al. 2005; Ho et al. 2007) and roots (Ho et al. 2007). The EST clones were also used as probes on cDNA microarrays (Low et al. 2006; Tee et al. 2013) and macroarrays (Lin et al. 2009; Beulé et al. 2011) to study gene expression.

One of the limitation of the EST approach is that it tends to tag medium or high abundant genes. Low abundant genes, which may have important roles in plant development, could be missed. At that time, two approaches were proposed to isolate low abundant genes, i.e. via saturation hybridization of genomic DNA to cDNAs (Weissman 1987) or through the construction of normalized cDNA libraries by reannealing cDNAs based on second-order kinetics (Soares et al. 1994; Bonaldo et al. 1996). The method described by Bonaldo et al. (1996) was used to construct an oil palm normalized library, and 237 non-redundant low abundant genes were identified (Chan et al. 2010).

The ESTs were also an important resource to identify markers for genetic mapping programmes, especially those that are linked to expressed genes. The clones were utilized as restriction fragment length polymorphism (RFLP) markers (Maizura et al. 2006; Singh et al. 2008a), while the sequences were mined for single-nucleotide polymorphisms (SNPs) and simple sequence repeats (SSRs) (Riju et al. 2007;

Low et al. 2008; Singh et al. 2008b; Ting et al. 2010). Another important source of markers was developed by Billotte et al. (2001) using SSR-enriched libraries. These SSR markers, together with amplified fragment length polymorphism (AFLP) markers, were used to identify additional markers flanking the *SHELL* gene (Billotte et al. 2005), the first oil palm monogenic trait mapped (Mayes et al. 1997). These markers made it possible to develop genetic maps and identify quantitative trait loci (QTLs) influencing callogenesis rate (Ting et al. 2006), fatty acid composition (Singh et al. 2009; Montoya et al. 2013), yield components (Billotte et al. 2010; Tisné et al. 2015), vegetative measurements (Billotte et al. 2010), and sex ratio and related traits (Ukoskit et al. 2014). The markers also provided an avenue for researchers to study genetic diversity of *E. guineensis* (Maizura et al. 2006; Singh et al. 2008b; Ting et al. 2010; Okoye et al. 2016).

To improve the repertoire of genes and markers for research, the Malaysian Palm Oil Board (MPOB) embarked on a small-scale genome sequencing project using the GeneThresher (GT) technology. The technology was selected then as it was still very costly to carry out whole-genome sequencing of a large plant genome. Using GT, the genic or hypomethylated regions of the oil palm genome were selectively sequenced (Budiman et al. 2005; Low et al. 2014). The two main objectives were to develop a new set of markers for genetic mapping and as a resource to identify genes of interest. To achieve this, libraries from seven *E. guineensis* and two *E. oleifera* palms were constructed and sequenced. The palms were selected to provide enough diversity of the planting materials in Malaysia for marker development. About 307,000 *E. guineensis* and ~155,000 *E. oleifera* reads were assembled into ~201,000 (45,370 contigs and 155,442 singletons) and ~111,000 (18,836 contigs and 92,446 singletons) unique sequences, respectively. A total of 33,752 SSRs, 40,820 SNPs, 242 transcription factors, 65 candidate resistance genes and 40 microRNAs (miRNAs) were identified (Low et al. 2014).

The GT SNP marker collection was used to develop a 4451 SNP array that was eventually used together with the genome sequence and breeding information to identify the *SHELL* (Singh et al. 2013a) and *VIRESCENS* (*VIR*) (Singh et al. 2014) genes. The *SHELL* gene is responsible for the three oil palm fruit forms, *dura*, *tenera* and *pisifera*, while the *VIR* gene controls the colour of the fruit exocarp, an important indicator of fruit ripeness. The GT markers, used together with other SSR and AFLP markers, were successfully deployed to identify 164 QTLs linked to oil yield component traits, such as average bunch weight, bunch number, fresh fruit bunch, oil to bunch and shell to fruit (Seng et al. 2016). In oil palm interspecific hybrid programmes, the SSRs and SNPs were used to develop high-density genetic maps (Ting et al. 2014), identify QTLs linked to fatty acid composition (Ting et al. 2016) and study marker inheritance in backcross two hybrid mapping populations (Zulkifli et al. 2014). The *E. oleifera* markers were used to study marker transferability between the South American oil palm, coconut and ornamental palms (Zaki et al. 2012). GT sequences were also used to develop SNP-based cleaved amplified polymorphic sequence (CAPS) markers for an oil palm genetic diversity study (Ong et al. 2015). A SNP marker associated with height increment was also identified (Ong et al. 2018). Table 7.1 lists a compilation of publications using markers from EST, SSR-enriched libraries and GT data for gene identification, genetic mapping and diversity studies.

7.3 Oil Palm Genome

The Algemene Vereniging van Rubberplanter ter Oostkust van Sumatra (*AVROS*) *pisifera* fruit form of *E. guineensis* was sequenced, assembled and published in 2013 using a combination of 454 pyrosequencing and bacterial artificial chromosome (BAC)-end sequences (Singh et al. 2013b). The advantage of using pyrosequencing is that it is more cost effective and has higher throughput compared to Sanger technology.

Table 7.1 Selected publications using markers from EST, SSR-enriched libraries and GT data

Title	Authors	Discoveries
Development, characterisation, and across-taxa utility of oil palm (*Elaeis guineensis* Jacq.) microsatellite markers	Billotte et al. (2001)	Development and utilization of SSRs from SSR-enriched libraries
Microsatellite-based high density linkage map in oil palm (*Elaeis guineensis* Jacq.)	Billotte et al. (2005)	SSR and AFLP high-density map with *SHELL* gene locus
Statistical mapping of quantitative trait loci controlling the time to first callusing in oil palm (*Elaeis guineensis* Jacq.) tissue culture	Ting et al. (2006)	QTLs associated with time to first callusing
Assessment of genetic diversity in oil palm (*Elaeis guineensis* Jacq.) using restriction fragment length polymorphism (RFLP)	Maizura et al. (2006)	Genetic diversity study of oil palm germplasm material from 11 African countries
Identification of cDNA-RFLP markers and their use for molecular mapping in oil palm (*Elaeis guineensis*)	Singh et al. (2008a)	QTL associated with oil to wet mesocarp (O/WM) content
Exploiting an oil palm EST database for the development of gene-derived SSR markers and their exploitation for assessment of genetic diversity	Singh et al. (2008b)	Genetic diversity study using EST-SSR markers
Mapping quantitative trait loci (QTLs) for fatty acid composition in an interspecific cross of oil palm	Singh et al. (2009)	QTLs associated with fatty acid composition
QTL detection by multi-parent linkage mapping in oil palm (*Elaeis guineensis* Jacq.)	Billotte et al. (2010)	Seventy-six QTLs associated with yield components and vegetative measurements
SSR mining in oil palm EST database: application in oil palm germplasm diversity studies	Ting et al. (2010)	Genetic diversity study of oil palm germplasm and transferability of *E. guineensis* EST-SSRs among palm species
Elaeis oleifera genomic-SSR markers: exploitation in oil palm germplasm diversity and cross-amplification in Arecaceae	Zaki et al. (2012)	Genetic diversity study using SSRs from *E. oleifera* genomic libraries
The oil palm *SHELL* gene controls oil yield and encodes a homologue of SEEDSTICK	Singh et al. (2013a)	Identification of the oil palm *SHELL* gene
Quantitative trait loci (QTLs) analysis of palm oil fatty acid composition in an interspecific pseudo-backcross from *Elaeis oleifera* (H.B.K.) Cortés and oil palm (*Elaeis guineensis* Jacq.)	Montoya et al. (2013)	Nineteen QTLs associated with fatty acid composition
Identification of QTLs associated with callogenesis and embryogenesis in oil palm using genetic linkage maps improved with SSR markers	Ting et al. (2013)	Two QTLs associated with callusing and embryogenesis rates
The oil palm *VIRESCENS* gene controls fruit colour and encodes a R2R3-MYB	Singh et al. 2014	Identification of the oil palm *VIR* gene
High density SNP and SSR-based genetic maps of two independent oil palm hybrids	Ting et al. (2014)	High-resolution genetic maps of an intraspecific cross (Deli *dura* and Yangambi *pisifera*) and an *E. oleifera* × *E. guineensis* (O×G) hybrid constructed using SNP and SSR markers
Oil palm (*Elaeis guineensis* Jacq.) linkage map, and quantitative trait locus analysis for sex ratio and related traits	Ukoskit et al. (2014)	Eight QTLs across six linkage groups associated with sex ratio and related traits

(continued)

Table 7.1 (continued)

Title	Authors	Discoveries
Inheritance of SSR and SNP loci in an oil palm interspecific hybrid backcross (BC2) population	Zulkifli et al. (2014)	Development of markers for interspecific hybrid backcross (BC2) population
A consensus linkage map of oil palm and a major QTL for stem height	Lee et al. (2015)	A major QTL for stem height
Development of SNP markers and their application for genetic diversity analysis in the oil palm (*Elaeis guineensis*)	Ong et al. (2015)	SNP-based CAPS markers for genetic diversity study of germplasm material from Angola
Mixed model approach for IBD-based QTL mapping in a complex oil palm pedigree	Tisné et al. (2015)	Eighteen QTL regions controlling production traits
Genetic relationships between elite oil palms from Nigeria and selected breeding and germplasm materials from Malaysia via simple sequence repeat (SSR) markers	Okoye et al. (2016)	Genetic diversity study of elite oil palm materials from Nigerian Institute for Oil Palm Research (NIFOR) and Malaysia
QTLs for oil yield components in an elite oil palm (*Elaeis guineensis*) cross	Seng et al. (2016)	QTLs associated with yield component traits
Fine-mapping and cross-validation of QTLs linked to fatty acid composition in multiple independent interspecific crosses of oil palm	Ting et al. (2016)	QTL linked to iodine value (IV) and fatty acid composition (FAC) identified in an O×G cross
Association of SNP markers with height increment in MPOB-Angolan natural oil palm populations	Ong et al. (2018)	SNP marker associated with height increment

454 also has longer reads compared to the Illumina technology, and this was deemed necessary as the oil palm is an ancient tetraploid that has a relatively large genome of ~1.8 Gb. Nevertheless, pyrosequencing has limitations in the detection of long homopolymeric repeats (Siqueira et al. 2012). When the oil palm genome programme was initiated, the assumption was that since oil palm has a relatively large genome, ~3 times larger than the closely related date palm which has a genome of ~605 Mb (Al-Mssallem et al. 2013), the repeat regions would be large. A pool of large linker libraries (0.5, 5, 13 and 20 kb) was constructed and sequenced to try to bridge over the repeat regions. This was not the case when a review of the initial sequencing results was done. Careful comparison of the assembled scaffold sequences to available oil palm BAC sequences that were manually finished showed that the gaps in the sequences were relatively small, with >80% of them being <3 kb (Fig. 7.2a). The profile of the repeat region lengths has since been confirmed using one of the latest oil palm genome builds available at MPOB. The *AVROS pisifera* Version 8 (P8) genome build, which has only 0.53% scaffold gap, showed that the majority of the repeats identified (~79%) were <2 kb (Fig. 7.2 b). The gap length profile made the team review its strategy to focus on smaller size paired-end and linker libraries (0.75, 1.5, 3 and 8 kb) (Fig. 7.3). Fibonacci numbered size libraries were used so that doublings of insert sizes or the merger of two fragments do not sum up to a larger sized library insert range. This issue is somewhat mitigated nowadays with the latest advancements of third-generation sequencing technologies, such as the single-molecule sequencing technologies developed by PacBio BioSciences (PacBio). PacBio technologies are able to generate much longer reads that are able to sequence through repeat regions (Rhoads and Au 2015).

With additional sequences from the smaller libraries, the N50 (minimal contig or scaffold length where 50% of the entire assembly is

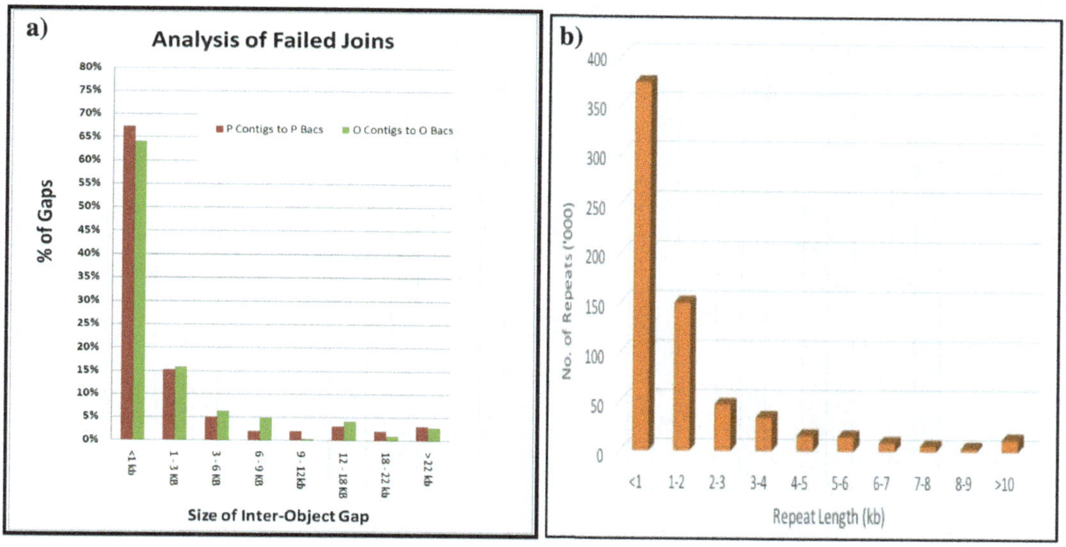

Fig. 7.2 Distribution of gaps and repeats in scaffold sequences. **a** Scaffold sequences from the *AVROS pisifera* P1 genome build were compared to manually finished BAC sequences that were at least 100 kb in length. The size of gaps was determined by the size of the respective regions in the BAC sequences. **b** The repeats in the *AVROS pisifera* P8 genome build were analysed using RepeatModeler, and the number of repeats identified was plotted according to size

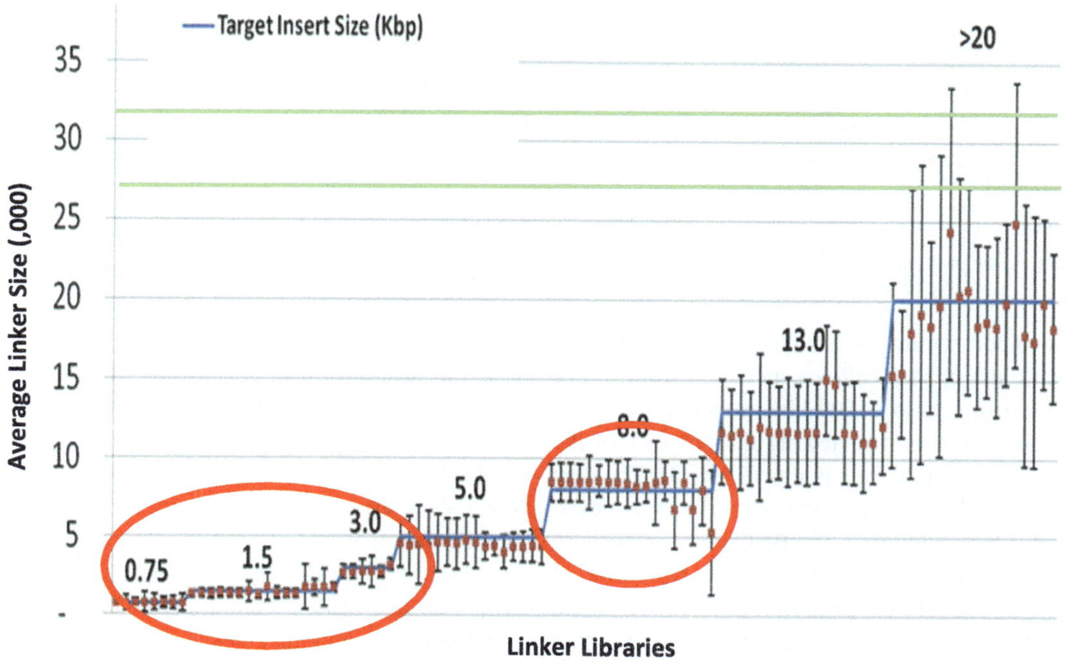

Fig. 7.3 Distribution of 454 linker libraries. The red circles are the additional libraries that were added after the gap analysis of the P1 build. The blue line is the target insert size of the library, while the red dots are the actual mean insert size in kilobase (*Source* Singh et al. 2013b)

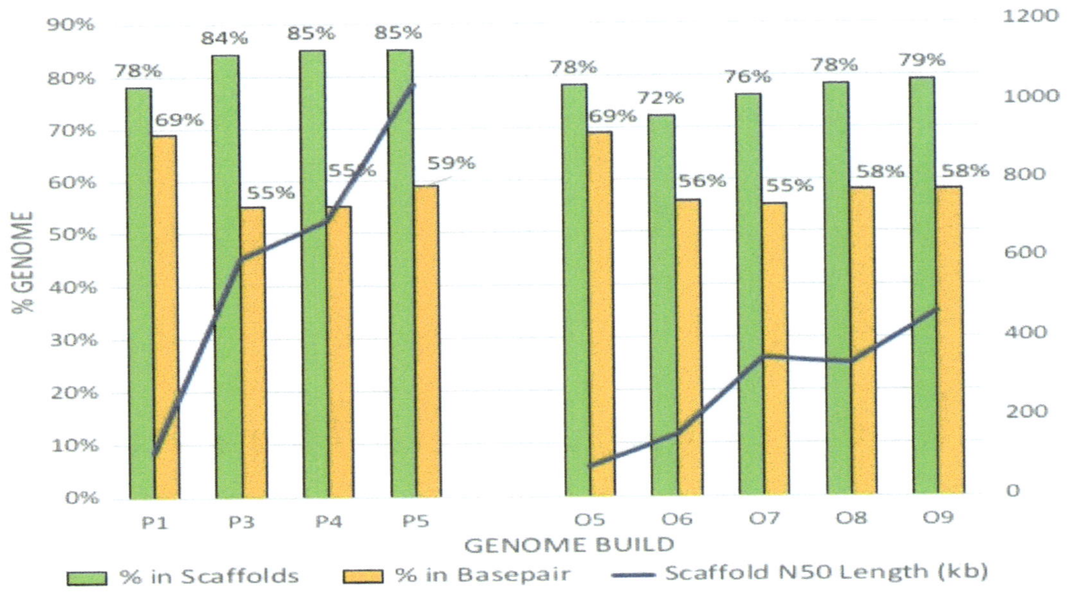

Fig. 7.4 Statistics of oil palm genome builds

Table 7.2 Statistics of P5 genome build

Genome	N	Size (Mb)	N50 (Kb)	Largest (Mb)	G + C Content (%)
P5 build contigs	1,309,411	1549	3.14	0.13	39
P5 build scaffolds	40,360	1535	1045	22.10	37
EG5 linked chromosomes	16	658	44,350	68.43	37
EG5 linked chromosomes + P5 unlinked	40,072	1535	1270	68.43	37
Scaffold gaps	166,221	478	–	–	–

Source Singh et al. (2013b)

contained within the smallest set of contigs or scaffolds) of the scaffolds increased from ~112 kb (version 1) to ~600 kb (version 3), showing the importance of reviewing the progress of the programme at all time points (Fig. 7.4). The next jump in the scaffold N50 length was achieved with the addition of BAC-end sequences, resulting in ~50% increase in the N50 scaffold length. The *AVROS pisifera* Version 5 (P5) build (Table 7.2), which had 40,360 scaffolds with an N50 size of 1.05 Mb, covered ~1.5 Gb of the total 1.8 Gb genome (Singh et al. 2013b). The P5 genome was released into the public domain via GenBank and the Genomsawit website (http://genomsawit.mpob.gov.my). Nevertheless, due to the use of linker libraries and BAC-end sequences, the scaffolds had ~31% gap. Using two genetic maps (P2 and T128), ~660 Mb of scaffolds in the P5 build were linked as pseudochromosomes (EG5), of which ~69% of the scaffolds in the pseudochromosomes were able to be oriented as it had at least two markers localized onto the scaffold. The remaining scaffolds with only one marker were not oriented. Also, it is important to note that since oil palm is a tree, the number of palms per mapping population is small. The genetic maps are good for the overall organization of the genome but might have some inconsistency at the local order. This is because falsely scored markers or incorrectly ordered markers have larger effects in smaller populations (Sim et al. 2012). The *E. oleifera* (Singh et al. 2013b) and *E. guineensis dura* (Jin et al. 2016) genome assemblies were also released. The *E. oleifera* genome was de novo assembled using Newbler (Singh et al. 2013b) and had a scaffold N50 of ~333 kb (Singh et al. 2013b; Filho et al. 2017). The *dura* genome that was assembled using a combination of de novo assembly and alignment to the *AVROS pisifera* EG5 reference genome had longer scaffolds, with scaffold N50 of ~750 kb (Jin et al. 2016).

Analysis of the *AVROS pisifera* genome shows that it has an estimated repeat content of ~57%, of which ~27% (282 Mb), ~47% (478 Mb) and ~26% (265 Mb) were independent genomic regions that had matches to repetitive sequence elements, scaffold gaps and the un-sequenced portions of the genome, respectively. The majority of the repeats observed were uncharacterized, with most of them (73%) being absent in the *E. oleifera* genome (Singh et al. 2013b), which has ~56% transposable elements in its sequence (Beulé et al. 2015). As expected, long terminal repeat (LTR) retroelements were mainly represented by Copia elements (Singh et al. 2013b; Beulé et al. 2015; Filho et al. 2017). These Copia elements had a frequency of 5:1 compared to the Gypsy LTR retroelements and were preferentially found in gene-rich regions (Beulé et al. 2015). This is in agreement with fluorescence in situ hybridization (FISH) results, which showed that Copia-like retroelements were dispersed throughout the genome (Castilho et al. 2000). Class II DNA transposons and other non-LTR retrotransposons were observed at much lower frequencies (Singh et al. 2013b; Filho et al. 2017).

Comparison of the *AVROS* chromosome sequences to each other and to the *E. oleifera* genome showed that the genomes were duplicated and that the duplications occurred before the divergence of the two oil palm species (Singh et al. 2013b), in line with earlier speculations that oil palm had possible tetraploid ancestry (Castilho et al. 2000; Singh et al. 2008b). This was supported by syntenic block analysis, which showed that oil palm and date palm shared two paleotetraploidy events, one (τ) occurring when monocots diverged from dicots while another (ρ) occurred only in palms (Jiao et al. 2014; McKain et al. 2016). Synonymous site substitution (Ks) rate of rice and oil palm paralogs formed in the shared τ whole-genome duplication (WGD) event shows that rice is evolving faster than oil palm. Its substitution rate is ~1.7 times larger than oil palm (Jiao et al. 2014). Molecular evolution is hypothesized to be effected by generation time, and since oil palm has a longer generation time compared to rice, this probably resulted in the lower molecular evolution rate (Low et al. 2017). Smith and Donoghue (2008) also found that palms had lower molecular evolution rates compared to other monocots and associated it with the shift to tree/scrub habit in palms. Using the Ks distribution of mode and estimated timeline of evolutionary history of banana (D'Hont et al. 2012) and palms (Singh et al. 2013b), the palm WGD ρ event was estimated to have occurred ~75 million years ago (MYA) (Jiao et al. 2014), close to the period (~60–70 MYA) where many independent WGD events occurred in plant lineages (Van de Peer et al. 2009). These polyploidization events coincide with the Cretaceous–Tertiary boundary where ~60% of plant species became extinct, suggesting that polyploidization may have helped improve the adaptability and survivability of several plant lineages (Fawcett et al. 2009).

Analysis of duplicated regions indicates that most of the oil palm centromeres arose after polyploidization, since these regions did not span putative pericentromeric regions. The analysis also revealed that oil palm chromosomes 2 and 14 were formed as a result of Robertsonian fusion of two date palm chromosomes, respectively (Singh et al. 2013b). This theory was supported by Mathew et al. (2014), who reported that date palm chromosomes 1 and 10 fused to form oil palm chromosome 2. Mapping of date palm and coconut markers to the oil palm genome sequence also showed that the three palms maintained long-range synteny despite the different genome sizes (Mathew et al. 2014).

7.4 Gene Models

With the availability of next-generation sequencing (NGS), the ability to sequence new genomes has been on the increase. This makes it more pertinent to annotate genomes by identifying and characterizing the genic regions accurately. This can be achieved using software that predicts the protein-coding genes based on Hidden Markov model (HMM) approaches (Stanke and Waack 2003; Majoros et al. 2004; Korf 2004; Ter-Hovhannisyan et al. 2008), support vector machine (Schweikert et al. 2009) or by combining predictions and evidence from multiple sources (Snyder and Stormo 1995; Ying et al. 1996). The initial gene models of the oil palm genome were predicted using a combination of Glimmer (Majoros et al. 2004) and SNAP (Korf 2004), which is based on Markov models. Glimmer's HMM was trained using exon boundaries of a selected set of mapped RNA-Seq sequences, while SNAP was trained using an initial set of oil palm genes predicted using rice HMM model. Since Glimmer produced better predictions, Glimmer predictions were selected to represent the locus when models from both software overlapped, resulting in a total of 158,946 gene models. After filtering sequences that have similarity to known retroelements, 34,802 genes with RefSeq support were identified (Table 7.3). This first set of gene models was useful in identifying genes involved in oil synthesis and carbohydrate metabolism. Interestingly, comparison of oil palm and date palm genes showed that even though both palms had similar number of triacylglycerol biosynthesis genes, sucrose degradation and oxidative pentose phosphate pathway genes were more highly represented in oil palm compared to date palm (Singh et al. 2013b). The gene models were also used to identify the oil palm *SHELL* (Singh et al. 2013a) and *VIR* genes (Singh et al. 2014). Nevertheless, Benchmarking Universal Single-Copy Orthologs (BUSCO) analysis, which is a set of near-universal single-copy orthologs used to provide quantitative assessments of the completeness of genome assemblies and gene sets (Simão et al. 2015), showed that the first set of P5 gene models had many fragmented genes. The gene models only had 60.35% of the *Embryophyta* BUSCO profiles, of which 68.9% of them were complete genes. The gene predictions from the *dura* genome were slightly better, having a total of 910 (63.2%) out of the 1440

Table 7.3 Comparison of gene models from the *pisifera* and *dura* genome builds

Parameter	EG5 genome build (Singh et al. 2013b)	EG5 genome build (Chan et al. 2017b)	*Dura* genome build (Jin et al. 2016)
Gene number	34,802	26,059[a]	36,105
Average length (bp)	1000	1237	3573
Gene density (gene/Mb)	22.67	16.98	21.23
Average exon per gene	–	5.4	3.7
BUSCO			
Complete and single copy	552	1031	687
Complete and duplicated	82	139	78
Fragmented	266	65	145
Missing	540	205	530
Total BUSCO profiles	1440	1440	1440

[a]High-quality gene models with RefSeq and transcriptome evidence

BUSCO profiles. A total of 145 BUSCO profiles in the *dura* gene models were fragmented (Chan et al. 2017b).

To overcome these limitations, a new pipeline had to be developed to improve the quality of the gene models. The predictions from Singh et al. (2013b) provided an insight into how to improve the process. Based on the two gene modellers used, Glimmer produced better predictions. One of the reasons for this was because Glimmer used the oil palm transcriptome data as its training set. The advantage of this is that the transcriptome data provide the closest set of gene sequences that reflects oil palm's codon usage bias. Codon usage bias or codon bias is the selective usage of three nucleotides (a triplet) to code for synonymous codons in a protein-coding gene. Organisms have codon bias that influences the frequency of which synonymous codon they prefer to use (Campbell and Gowri 1990; Hershberg and Petrov 2008; Galperin and Cochrane 2009; Iriarte et al. 2013; Soltani et al. 2014). One of these differences is that the optimal codons for non-grass monocots have a G/C preference in the third codon position. Non-grass monocots also have a lower preference for using XUA codons [AUA (ILE), UUA (LEU), CUA (LEU) and GUA (VAL)] in their protein-coding genes (Mazumdar et al. 2017). Also, date and oil palm, which belong to the non-grass monocot species, have unimodal GC_3 (guanidine and cytosine content of the third or wobble position of the codon) distributions with a long tail that is skewed towards high values (Clément et al. 2015), while grass monocots have bimodal GC_3 distribution (Clément et al. 2015; Mazumdar et al. 2017). *Arabidopsis thaliana,* on the other hand, has a unimodal GC_3 distribution (Tatarinova et al. 2013). These differences in codon bias and the atypical distribution of nucleotides in oil palm would impact the performance of gene model predictions.

To overcome this, a new pipeline, Seqping, was developed to automate and train species-specific HMMs using transcriptome data for more accurate gene predictions (Chan et al. 2017a). Seqping is a pipeline that is based on MAKER2 (Holt and Yandell 2011), and it uses the output of three modellers, SNAP (Korf 2004), AUGUSTUS (Stanke et al. 2008) and GlimmerHMM (Majoros et al. 2004) as part of its input data. The pipeline was validated using *A. thaliana* and rice genomes, where at least 95% of the Plantae BUSCO profiles were predicted (Chan et al. 2017a). Seqping improves prediction accuracy by selecting a set of putative full-length transcripts from the transcriptome data to develop species-specific HMMs. This is especially important for non-model genomes that do not have a collection of full-length genes or "gold standard gene models" to use for the training process. The process starts with the prediction of open reading frames (ORFs) from the assembled transcripts and comparing the ORFs to full-length coding sequences (CDSs) from GenBank. Once the ORFs are selected, the sequences are clustered and filtered for known repeats. The longest ORF for each cluster is selected as part of the full-length transcript training set. The reason for this is to develop a set of genes that represents as wide a range of genes as possible, while minimizing over-representation of a particular type or class of genes. This is to reduce the possibility of developing HMMs that would be trained to look for the over-represented genes, resulting in lower-quality predictions of other genes. This is important as non-grass monocot genes have very high levels of GC content heterogeneity, ranging from 20 to 80% (Mazumdar et al. 2017). This heterogeneity has also been observed in grass monocots (Tatarinova et al. 2010). Large differences in rice gene codon usage patterns correlate with its heterogeneity in nucleotide content (Wang and Hickey 2007). The reduction in the number of genes in the training set also reduces the compute power required for training.

The training set is used to develop individual species-specific HMMs for SNAP, AUGUSTUS and GlimmerHMM, which are subsequently used together with the genome sequence to predict the coding regions. The three sets of gene models are combined using MAKER2. MAKER2 is a multi-threaded and parallelized genome annotation pipeline that performs de novo annotation by integrating multiple ab initio prediction tools and

RNA-Seq data to improve the quality of the gene models predicted. It also adapted the Annotation Edit Distance (AED) measure developed by Eilbeck et al. (2009) to provide a quality control metric for each gene prediction. The AED score measures the conformity between the predicted gene and its supporting evidence. AED scores range from 0 to 1. Gene models with low AED scores have better agreement with the provided evidence support (Holt and Yandell 2011). The final steps in Seqping produce a set of gene models that have been filtered for sequences that are short or have significant similarity ($1e^{-10}$) to TIGR plant repeat database (Ouyang and Buell 2004), GIRI Repbase (Jurka et al. 2005) and Gypsy Database (Llorens et al. 2011).

For the oil palm P5 genome build, a computational framework that uses two independent gene prediction pipelines, Seqping and Fgenesh++ (Solovyev et al. 2006), was used to develop an improved set of gene models. The framework predicted 26,059 high-quality gene models with transcriptome and RefSeq support (Table 7.3). Assessment of the gene models using BUSCO showed that 90.44% of 429 eukaryotic and 85.76% of the 1440 *Embryophyta* BUSCO profiles were identified. A total of 81.25% of the *Embryophyta* profiles were present as complete BUSCO genes (Chan et al. 2017b). The new prediction had 51.20% (1170 vs. 599) more complete, 24.07% (65 vs. 270) less fragmented and 35.90% (205 vs. 571) less missing genes compared to the original gene predictions released by Singh et al. (2013b). The improvement was a result of using integrated gene prediction pipelines without any change in the P5 genome build. This shows the advantage of using gene prediction pipelines that provide the option to generate species-specific HMMs for gene prediction.

The improved P5 gene models (Chan et al. 2017b) were analysed to identify GC_3-rich genes, which had a characteristic increase in GC_3 from the 5' to 3' end of the gene. The analysis showed that oil palm GC_3-rich and GC_3-poor genes had an average length of ~1.9 kb and ~13 kb, respectively. This is in agreement with the results from other studies which showed that GC_3-rich genes tend to be shorter than GC_3-poor genes (Alexandrov et al. 2009; Tatarinova et al. 2013). Gene ontology analysis showed that the GC_3-rich and GC_3-poor genes were enriched for stress-related and housekeeping functions, respectively (Chan et al. 2017b). Sundararajan et al. (2016) also reported that high GC_3 genes were more prevalent in stress responses and immune systems, while the low GC_3 genes were highly represented in housekeeping or generalized functions, such as in cellular and metabolic processes, protein metabolism, biosynthesis pathways, transcription, transcription regulation and transport. The gene set also contained 3658 intronless genes with an average CDS of 924 nt, shorter than the average length of multi-exonic CDS (1289 nt). Interestingly, the intronless genes represent 51% of the GC_3-rich genes. This is similar to corn where the GC3-rich genes were also highly enriched with intronless genes (Alexandrov et al. 2009).

Interestingly, GC_3-rich genes tend to be rapidly evolving, have more variable expression and frequently lack paralogs (Tatarinova et al. 2010). This is probably due to the fact that GC_3 levels are highly correlated with double-strand break (DSB) hotspot motifs, providing a possible mechanism for the high evolutionary rates of these genes. GC_1- and GC_2-rich genes also have similar correlations. Although only one quarter of genes were classified in the high GC_x category, 70% of all genic DSB hotspots were found in them (Sundararajan et al. 2016). Tatarinova et al. (2010) suggested that GC_3-rich genes are maintained because of their optimized codon usage and additional methylation targets for transcriptional regulation. Nevertheless, this forms a paradox since GC_3 genes are negatively correlated with methylation (Tatarinova et al. 2013; Elhaik et al. 2014). Tatarinova et al. (2013) suggested that this could be because ubiquitously expressed GC_3-poor genes use methylation as one of the means to maintain broad expression while hypomethylated intronless genes are regulated via a different mechanism. CpG dinucleotide sites in GC_3-rich genes may be maintained to allow phenotypic plasticity.

Gene models from oil palm genomes were also used to identify resistance and fatty acid biosynthesis genes. The collection of resistance genes is an important resource for the study of infection and host–pathogen interactions. Jin et al. (2016) identified 566 candidate resistance (R) genes. Using a different approach, Chan et al. (2017a) used homology and multiple sequence alignment of oil palm genes to known R genes and characterization of domain structure to identify 210 candidate R genes with RefSeq and transcriptome evidence. These genes were categorized into six classes. Chan et al. (2017a) reported that the coiled-coil (CC)–nucleotide-binding site (NBS)–leucine-rich repeat (LRR) or CNL class R genes belonged to the largest class (141 CNL genes) of oil palm R genes identified. The CNL class is part of the NBS-LRR family, which is the biggest class of R genes in monocots. The family contains a conserved NBS domain that is involved in ATP binding and hydrolysis, while the LRR domain is required for protein–protein interactions. The amino-terminal domains of the LRR modulate activation, while the carboxyl-terminal forms the platform for upstream activators (Belkhadir et al. 2004). NBS-LRR family genes tend to cluster in the genome (McHale et al. 2006). Of the 101 CNL genes successfully placed on the oil palm EG5 pseudochromosomes, ~60% (62) formed 23 clusters (Chan et al. 2017b). Using NBS-LRR genes in sorghum, rice, maize and *Brachypodium*, Yang and Wang (2016) found that local duplication, especially in clusters, is the major source of evolution for NBS-LRR genes in grasses. These clusters form a reservoir of genetic variation, found mainly on the LRR domain for plants to counter pathogen attacks. Using the Ka/Ks (non-synonymous to synonymous site substitution) ratio, Jin et al. (2016) found that there was positive selection of the oil palm R genes.

As for the genes involved in fatty acid biosynthesis (FAB), 42 key genes were identified and categorized based on the conserved catalytic residues and motifs of their corresponding amino acid sequences. The analysis showed that oil palm contains both the multifunctional (2 copies) and multi-subunit (CT [3 copies], BCCP [2 copies], BC [2 copies]) forms of acetyl-CoA carboxylase (ACCase) genes (Chan et al. 2017b). This agrees with Alban et al. (2000) who reported the existence of ACCases in two distinct forms in plants. In oil palm, Omar et al. (2008) also identified two distinct forms of ACCase. ACCase catalyses the first building block for FAB by converting acetyl-CoA to malonyl-CoA. The malonyl-CoA is continuously extended by two-carbon acetyl group until palmitoyl-ACP (C16:0-ACP) is formed. The conversion of C16:0-ACP to C18:0-ACP (stearoyl-ACP) by β-ketoacyl-ACP synthase II (encoded by FABF) is the rate-limiting step for oil palm fatty acid synthesis. The bottleneck in the conversion of C16:0-ACP to C18:0-ACP, coupled with an active palmitoyl-ACP thioesterase that converts C16:0-ACP to C16:0 (palmitic acid), results in chain termination of de novo fatty acid synthesis and the channelling of palmitic acid towards triglyceride synthesis (Sambanthamurthi et al. 1999; Abrizah et al. 2000), explaining why oil palm accumulates 39.2–45.8% palmitic acid. Overexpressing FABF would increase oleic content of palm oil via the enhanced production of stearoyl-ACP stearate, which would be efficiently converted to oleoyl-ACP (C18:1-ACP) by stearoyl-ACP desaturase. C18:1-ACP would subsequently be hydrolysed to oleic acid which would then enter the triglyceride synthesis pathway. The gene model analysis also showed that the oil palm genome contains six copies of FAB2, which encode stearoyl-ACP desaturase, making it the highest copy number gene among all the FAB genes. The gene determines the ratio of saturated to unsaturated C18 fatty acid. Chan et al. (2017a) showed the presence of two copies of FATA (thioesterases that prefer unsaturated) and four copies of FATB (thioesterases that prefer saturated) in the oil palm genome. The analysis was further improved by Rosli et al. (2018), where the four-FATB gene classification was further refined to Class I (3 genes) and II (1 gene). Class I predominantly uses C14 and C16 substrates, while Class II has a broader base substrate range, with major activity towards C8 and C12 substrates.

7.5 Application of High-Throughput Technologies and Genome Data for Oil Palm Research

The availability of high-throughput technologies for gene expression studies and the genome data provided greater opportunities to improve our understanding and increase the diversity of oil palm research, ranging from programmes to identify and classify genes or pathways of interest (Chan et al. 2017b; Rosli et al. 2018) to large-scale gene expression studies (Table 7.4). The availability of high-throughput technologies also resulted in the design of more comprehensive experiments that led to the identification of QTLs (Pootakham et al. 2015; Bai et al. 2017) and markers (Teh et al. 2016) associated with traits of interest, and the development of predictive models for genomics selection (Kwong et al. 2017). The major breakthroughs include the identification of the genes that control fruit form (Singh et al. 2013a), exocarp colour (Singh et al. 2014) and the mantled somaclonal variant (Ong-Abdullah et al. 2015). Molecular diagnostic assays have been developed for these genes, and these will revolutionize the way the oil palm industry produces commercial planting materials. The summary of the publications using high-throughput technologies is available in Tables 7.4 and 7.5.

7.6 *SHELL* Gene

The *SHELL* gene controls the most important oil palm monogenic trait that is responsible for the three fruit forms, i.e. *dura*, *tenera* and *pisifera* (Fig. 7.5). Thin-shell *tenera* has a 0.5–3 mm thick endocarp and mesocarp equivalent to 60–95% of the fruit weight, while *dura* has a think endocarp (2–8 mm) and a medium size mesocarp (35–55% of fruit weight). *Pisifera* has no endocarp and a small pea-like kernel (Latiff 2000), and is predominantly female sterile. The heterozygous *tenera* (Sh^+ sh^-) is a hybrid of *dura* (Sh^+ Sh^+) and *pisifera* (sh^- sh^-) (Kushairi and Rajanaidu 2000). *Dura* and *pisifera* are normally used as the female and male parents, respectively. Although this co-dominant monogenic inheritance of *SHELL* was first discovered in Congo by Beirnaert and Vanderweyen (1941), it took another half-century for the genetic locus of the *SHELL* gene to be identified by Mayes et al. (1997). Additional attempts were made to narrow the region between the markers and the *SHELL* gene (Moretzsohn et al. 2000; Billotte et al. 2005), but the gene remained elusive. In 2013, Singh et al. (2013a) were able to identify the *SHELL* gene by using a combination of breeding information, genetic mapping and homozygosity mapping. The authors used five generations of breeding data to identify individual palms that have a single *AVROS* palm contributing to the *SHELL* gene locus. The idea was to generate a polymorphism map where multiple individuals contributed polymorphisms to all the locations in the genome except for the *SHELL* locus which is homozygous (single contributor). Fourteen *pisifera* palms were sequenced to a depth of 20× genome coverage, while another 29 *pisifera* palms were sequenced to 40× coverage. The sequences were mapped to the P5 genome. A homozygous region of ~200 kb was identified in pseudochromosome 2. This was compared to the T128 genetic map that segregates for the *SHELL* gene. The gene locus was identified on the 18.9 centimorgan (cM) position of linkage group 7. Of the 30 genes found in the homozygosity region, only five were fully homozygous, of which only one of the homozygous genes (homologue of *SEEDSTICK*) was in the genetic interval containing the *SHELL* locus (Singh et al. 2013a).

Sequencing of all the exons in the five genes identified two independent SNPs in exon 1 of the *SEEDSTICK* homologue gene that tracked with the *SHELL* phenotype. The wild-type allele was named $Sh^{DeliDura}$, while the allele from the *AVROS pisifera* was named sh^{AVROS}. The third allele, named sh^{MPOB}, was identified from the

Table 7.4 Summary of selected oil palm gene expression studies using high-throughput technologies

Title	Authors	Discoveries
Comparative transcriptome and metabolite analysis of oil palm and date palm mesocarp that differ dramatically in carbon partitioning	Bourgis et al. (2011)	Analysis showed that most enzymes of the triacylglycerol assembly and glycolysis were expressed at similar levels in oil palm and date palm, indicating that the major control of oil accumulation in oil palm is at the synthesis of fatty acids and supply of pyruvate in the plastid
Regulatory mechanisms underlying oil palm fruit mesocarp maturation, ripening, and functional specialization in lipid and carotenoid metabolism	Tranbarger et al. (2011)	Characterized key regulatory steps of pathways and genes linked to fruit maturation and ripening
Transcriptome of oil palm (*Elaeis guineensis* Jacq.) roots treated with *Ganoderma boninense*	Tee et al. (2013)	Identified a candidate isoflavone reductase gene that has the potential to be developed as an expression marker for oil palm basal stem rot disease
Transcriptome analysis of normal and mantled developing oil palm flower and fruit	Shearman et al. (2013)	Differential expressed genes between normal and mantled samples identified. The authors suggest that the mantled phenotype is a collection of phenotypes that after undergoing selection at plantlet stage, results in an apparent mantled phenotype. A previously identified potential marker for embryogenic callus was also found to be a potential marker for the mantled phenotype
Comparative transcriptome analysis of three oil palm fruit and seed tissues that differ in oil content and fatty acid composition	Dussert et al. (2013)	Two of the three paralogs of WRINKLED1 (WRI1), *EgWRI1-1* and *EgWRI1-2* were massively expressed in the mesocarp and endosperm during oil deposition. Expression of fatty acid synthesis genes correlated with WRI1 expression level and oil content
De novo transcriptome analyses of host-fungal interactions in oil palm (*Elaeis guineensis* Jacq.)	Ho et al. (2016)	Defence mechanisms of oil palm roots during *G. boninense* infection elucidated. Molecular interactions between oil palm and the biocontrol fungus; *Trichoderma harzianum* was also described
Differential gene expression at different stages of mesocarp development in high and low-yielding oil palm	Wong et al. (2017)	Down-regulation of sucrose metabolism-related genes in high-yielding compared to lower-yielding palms resulted in a higher carbon flux for up-regulated genes in glycolysis, tricarboxylic acid (TCA) and fatty acid biosynthesis, leading to enhanced oil production
Expression of microRNAs during female inflorescence development in African oil palm (*Elaeis guineensis* Jacq.)	Ho et al. (2017)	Differential gene expression of 18 orthologous miRNA families and their targets, and 15 putative oil palm-specific miRNAs were identified
Transcriptome analysis of cell wall and NAC domain transcription factor genes during *Elaeis guineensis* fruit ripening: Evidence for widespread conservation within monocot and eudicot lineages	Tranbarger et al. (2017)	Gene expression profiling of cell wall-related transcripts and NAC transcription factors during ripening suggests that ripening of monocotyledonous and eudicotyledonous lineages are regulated by evolutionarily conserved molecular physiological processes
Transcriptome and functional analysis reveals hybrid vigor for oil biosynthesis in oil palm	Jin et al. (2017)	*EgWRI1* identified as an important gene that contributes to hybrid vigour in lipid biosynthesis of oil palm
Comparative transcriptome analysis of oil palm flowers reveals an EAR-motif containing R2R3-MYB that modulates phenylpropene biosynthesis	Li et al. (2017)	Characterization of an EAR-motif-containing R2R3-MYB that modulates the metabolic flux of core phenylpropanoid pathway

(continued)

Table 7.4 (continued)

Title	Authors	Discoveries
Transcriptional response of oil palm (*Elaeis guineensis* Jacq.) inoculated simultaneously with both *Ganoderma boninense* and *Trichoderma harzianum*	Ho et al. (2018)	Transcriptome of oil palm roots treated with both *T. harzianum* and *G. boninense* revealed biocontrol mechanisms involving *T. harzianum* against *G. boninense* infection
Comparative genomic and transcriptomic analysis of selected fatty acid biosynthesis genes and CNL disease resistance genes in oil palm	Rosli et al. (2018)	Detailed classification of oil palm acyl-acyl carrier protein (ACP) thioesterases (FAT), stearoyl ACP desaturase (SAD) and resistance genes

Table 7.5 Selected publications on genetic mapping and genome-wide association study using NGS technologies

Title	Authors	Discoveries
Genome-wide SNP discovery and identification of QTL associated with agronomic traits in oil palm using genotyping-by-sequencing (GBS)	Pootakham et al. (2015)	Three QTLs affecting trunk height and one QTL associated with fruit bunch weight
Genome-wide association study identifies three key loci for high mesocarp oil content in perennial crop oil palm	Teh et al. (2016)	SNPs associated with oil to dry mesocarp yield trait
Genome-wide identification of markers for selecting higher oil content in oil palm	Bai et al. (2017)	Genotyping by sequencing (GBS) markers closely linked to four QTLs for oil content
Evaluation of methods and marker systems in genomic selection of oil palm (*Elaeis guineensis* Jacq.)	Kwong et al. (2017)	Predictive models for genomic selection of yield-related traits evaluated
Developing genome-wide SNPs and constructing an ultrahigh-density linkage map in oil palm	Bai et al. (2018)	Constructed an ultra-high-density linkage map for African oil palm
High density SNP and DArT-based genetic linkage maps of two closely related oil palm populations	Gan et al. (2018)	Development and integration of genetic maps using Diversity Arrays Technology (DArT) and SSR markers

germplasm samples. sh^{AVROS} was found in the Angola, Congo and Tanzania germplasm materials, while sh^{MPOB} occurred in Nigeria and Angola. Homozygous Sh^{DeliDura} ($Sh^{\mathrm{DeliDura}}/Sh^{\mathrm{DeliDura}}$) results in the formation of *dura* palms, while a combination of the wild-type allele with the recessive allele (sh^{MPOB} or sh^{AVROS}) produces *tenera* palms ($Sh^{\mathrm{DeliDura}}/sh^{\mathrm{MPOB}}$ or $Sh^{\mathrm{DeliDura}/sh\mathrm{AVROS}}$). All combinations of the recessive alleles ($sh^{\mathrm{AVROS}}/sh^{\mathrm{AVROS}}$, $sh^{\mathrm{AVROS}}/sh^{\mathrm{MPOB}}$ or $sh^{\mathrm{MPOB}}/sh^{\mathrm{MPOB}}$) result in *pisifera* palms (Singh et al. 2013a). The occurrence of heteroallelic *pisifera* with both the sh^{AVROS} and sh^{MPOB} alleles, as well as the heteroallelic *tenera* with the wild-type allele was supported by Teh et al. (2017). The authors reported that palms with heteroallelic recessive alleles produce *pisifera* while heteroallelic Sh^{DeliDura} with either the sh^{AVROS} or the sh^{MPOB} gave rise to *tenera* palms. Further investigations of the *SHELL* gene in the commercial materials in Malaysia resulted in the identification of three additional independent mutations (sh^{MPOB2}, sh^{MPOB3} and sh^{MPOB4}) that can result in *tenera* palms (Ooi et al. 2016).

A diagnostic assay was developed to take into consideration all the combinations of alleles that were discovered (Singh et al. 2015; Low et al. 2016; Ooi et al. 2016). The importance of this is that planting materials can now be selected at seed or nursery stage, saving time, cost and limited land resources. Commercial planters can now select to plant only *tenera* palms, which

Fig. 7.5 Cross section of oil palm fruits. **a** *dura* (thick shell), **b** *pisifera* (shell-less) and **c** *tenera* (thin shell). *Tenera* is a hybrid of *dura* and *pisifera*

have higher oil yield. Prior to this, planters could only identify *dura* or *pisifera* contaminations in the field 3–4 years after planting when the trees fruit. Even though most oil palm breeders and seed producers opine that careful control of pollination would limit dura contamination in D×P crosses to less than 1%, contamination in the 1980s was much higher. Two reports showed that dura contamination at two plantation companies was at 10–20% and in individual plantings, up to 45%. This was due to the introduction of the pollinating weevil, *Elaeidobius kamerunicus* in 1982 (Corley 2005). Uncontrolled wind and insect pollination, especially in plantations that do not have stringent quality control measures, were also reported as the possible causes of contamination of ~10% in selective breeding trials (Chin 1995). With improvements in control measures, the contamination rate was reduced in the 1990s. Comparison of seed sources in 1991 showed that the contamination level was reduced to below 2% (Rao and Kushairi 1999). Nevertheless, this is based on a sampling of a small portion of the trees planted. With the availability of the SHELL diagnostic assay, a more comprehensive survey was carried out. Screening of 57 nurseries and independent planting sites in Malaysia showed that the non-*tenera* contamination rate was ~11%. Eight of the 13 regions (3–6 sites per region) surveyed had an unadjusted non-*tenera* contamination rate of >10% (Ooi et al. 2016). One of the reasons for this is that there can be problems in certain batches of seeds (Low et al. 2016). This is because *dura* palms can produce rudimentary anthers which can result in accidental self-fertilization and subsequent *dura* contamination (Corley 2005).

The SHELL diagnostic assay is also important in breeding programmes. *Pisifera* and *tenera* palms from *tenera* × *pisifera* or *tenera* × *tenera* crossing programmes can be identified in the nursery and planted separately, improving the subsequent management and evaluation of the palms (Singh et al. 2013a; Syed Alwee et al. 2014). Planting of *pisifera* palms at higher density also encourages the formation of male flowers, increasing the production of pollen (Devuyst 1953). The confirmation of fruit form of parental lines also ensures that only progenies with the desired fruit form are produced. This is especially important when working with germplasm material, as thick-shell *tenera* can sometimes be mistaken as *dura* (Low et al. 2016). Testing of parental lines for commercial seed production and their subsequent progenies will also provide a guarantee of the quality of the planting materials used.

7.7 *VIRESCENS* Gene

Oil palm fruits are divided into two phenotypes for fruit colour, namely *Nigrescens* (NIG) and VIR. NIG, which is found in the majority of commercial oil palm plantings in South-East Asia, is a recessive trait that produces fruits that are deep violet to black at the apex and pale yellow at the base when unripe, and red or purplish-black at the apex and red at the base when ripe. VIR palms produce unripe green fruits and reddish-orange fruits when ripe (Hartley 1988) (Fig. 7.6). The negligible change in the NIG fruit colour during ripening makes it difficult to determine the ripeness of the fruit, resulting in the harvesting of under- or overripe bunches. Unripe bunches have not reached full stage of maturity and therefore have lower oil content, while overripe fruits result in higher labour cost for loose fruit collection as the fruits will abscise from the bunch during harvesting. The free fatty acid content would also increase significantly in overripe fruits, resulting in deterioration of oil quality. In the field, harvesters rely on the number of loose NIG fruits on the ground as an indicator of fruit ripeness. The use of VIR palms is advantageous to the industry as the striking change in colour from unripe to ripe fruits makes it easy to select ripe fruits for harvesting, giving the highest oil yield, while maintaining most of the fruits on to the bunch. Quality control of the harvested fruits at collection centres (Fig. 7.7) and the mill can also be carried out, as the ripe and unripe bunches can be easily identified, increasing oil extraction rates at the mills. Interestingly, although NIG palms are the most common palms available in natural grooves, the *VIR* gene is the dominant trait. This was probably because the orange fruits were used only for ceremonial purposes and not as food (Royal Botanic Gardens Kew 1909; Zeven 1967).

The locus of the gene was identified on linkage group 1 (chromosome 1) using the selfed-T128 mapping population that was used to identify the *SHELL* gene. The locus was most tightly linked to an RFLP probe (MET16) that was mapped 0.6 cM distance from the gene and was able to distinguish NIG and VIR at 95% accuracy. Using a minimal tiling path of BAC contigs and additional SNP markers, the *VIR* locus was narrowed down to an interval containing four genes. Analysis of the mutations in these genes showed that only one gene, a R2R3-MYB transcription factor, had mutations that tracked with the trait. Subsequent screening of germplasm samples resulted in the identification of an additional four mutations. The five independent mutations, identified in exon 3, result in premature termination of the amino acid sequence (Singh et al. 2014). This is similar to the *C1* gene in maize, where the C-terminal truncations result in dominant-negative allelic forms (McClintock 1951). The oil palm R2R3-MYB gene is closely related to Lilium LhMYB12 and has significant similarity to *Arabidopsis* production of Anthocyanin Pigment 1 (PAP1) and AtMYB113 (Singh et al. 2014). In Asiatic hybrid lilies, LhMYB12 is positively correlated with tepal anthocyanin pigmentation (Yamagishi et al. 2012). Overexpression of PAP1 in *Arabidopsis* and tobacco results in enhanced pigmentation (Borevitz et al. 2000), while overexpression of AtMYB113 up-regulates the anthocyanin pathway (Gonzalez et al. 2008).

Transcriptome analysis of the oil palm NIG and VIR fruits at 8 weeks after anthesis

Fig. 7.6 Virescens and Nigrescens fruits. Ripe and unripe **a** VIR and **b** NIG bunches and spikelets

Fig. 7.7 Quality control of harvested fruits. Unripe or partially ripe VIR fruits can be easily identified

(WAA) showed that the oil palm R2R3-MYB gene and most of the anthocyanin pathways genes starting from the trans-cinnamate 4-monooxygenase (C4H) step had higher expression in NIG compared to VIR fruits, indicating that the truncated *VIR* mutations inhibited the MYB-regulated genes. This was confirmed by ultraviolet–visible spectrophotometric and high-performance liquid chromatography (HPLC) analyses of acidified methanol extracts of exocarp, which shows that anthocyanin was present in NIG but absent in VIR fruits (Singh et al. 2014). The results show that in the absence of anthocyanin, we are able to see the underlying green colour of chlorophyll and orange-coloured carotenoids in unripe and ripe VIR fruits, respectively, which are masked by the anthocyanin in NIG fruits. With the identification of the *VIR* gene, a diagnostic assay has been developed to allow materials to be screened for their fruit exocarp colour as early as in seed or nursery stage. Being a dominant trait, both homozygous and heterozygous *VIR* alleles will result in VIR palms. However, if heterozygous VIR palms were to be used in breeding or commercial planting programmes, the subsequent progenies will segregate for the trait to produce both VIR and NIG palms. Therefore, the importance of the diagnostic assay would be to allow for the selection of high-yielding homozygous VIR parental palms that do not segregate for the trait in crossing programmes (Low et al. 2016). The biggest impact of the discovery would be the selection of homozygous *VIR* gene in non-abscising or delayed abscising genotypes that were previously observed to exist in oil palm fields by Osborne et al. (1992). Non-abscising VIR palms will improve yield and efficiency, as only ripe fruits will be harvested and processed. It will also reduce the need for loose fruit collection.

7.8 *MANTLED* Gene

Somatic embryogenesis is a unique process that demonstrates the plasticity of plant development. The process involves the production of morphologically and developmentally normal embryos from somatic cells that have the ability to develop into whole plants (Ong-Abdullah and Ooi 2006). This process can be induced under certain in vitro conditions using appropriate competent cells and hormones (Gaj et al. 2005) and encapsulates a complex myriad of events that have been the focus of numerous studies (Ikeda et al. 2006). In oil palm, tissue culture is predominantly initiated from oil palm young leaves in callus induction media. The callus, which is nodular in appearance, forms along the cut edges where the veins are exposed (Rohani et al. 2000). Some of the callus would remain compact and nodular, and undergo embryogenesis (Kanchanapoom and Domyoas 1999; Rohani et al. 2000). The process would eventually lead to somatic embryo formation and maturation, shoot regeneration, rooting and finally the recovery of new viable plantlets. Tissue culture is an important development in the oil palm industry, as it provides a vegetative means to shortcut the long breeding cycle of oil palm and produce elite planting materials. Clonal true-to-type elite *tenera* hybrids can increase yields by 20–30% (Corley and Law 1997).

The first successful reports on clonal propagation of oil palm were described by Raberchault et al. (1970) and Jones (1974). After these discoveries, the industry started investing in tissue culture. By the 1980s, 10 oil palm tissue culture laboratories were established (Wooi 1990). The enthusiasm, however, was dampened when Tan Yap Pau from United Plantation (Ong-Abdullah et al. 2015) first reported the occurrence of mantled clonal palms (Fig. 7.8). This was subsequently documented by Corley et al. (1986). The phenomenon known as mantling involved the feminization of male parts in male and female flowers to form pseudocarpel structures (Corley et al. 1986; Rival 2000; Rival et al. 2001). Mantling in male flowers results in the absence of pollen formation, while in female flowers, the mantled fruits that formed have lower yields. In cases of severe mantling, the fruits are sterile and aborted. Abnormalities of both flowers negatively affect the overall yields of plantations. Nevertheless, depending on the severity of the

Fig. 7.8 Normal and abnormal mantled palms. Oil palm trees, bunch, spikelet and cross and longitudinal sections of fruits from **a** normal, **b** fertile mantled and **c** sterile (parthenocarpic) mantled palms. The arrows show supernumerary carpels

abnormality, mantled palms have been able to revert towards normal phenotype over time (Durand-Gasselin et al. 1990).

A concerted effort was initiated by the Palm Oil Research Institute of Malaysia (now MPOB) and the oil palm industry to work together to understand the molecular mechanisms that cause mantling. Earlier efforts failed to identify the underlying differences between abnormal and normal palms using anatomical, cytogenetics, hormonal and isozyme studies (Ong-Abdullah et al. 2016). Nevertheless, the reversion (Durand-Gasselin et al. 1990) and decrease in DNA methylation (Jaligot et al. 2000; Matthes et al. 2001; Kubis et al. 2003) in mantled palms indicated a possible link between epigenetic change and mantling. Adam et al. (2005) suggested that the phenomenon may be caused by modifications of B-function MADS box genes, as mantled palms resembled B-function MADS box gene mutants in monocot and dicot species. Jaligot et al. (2014) studied the relationship of a B-type MADS box gene, *DEFICIENS* to the mantled floral phenotype using the methylation and expression data of the gene and two retrotransposons, a gypsy retrotransposon (*Koala*) in intron 5 and a Copia retrotransposon (*Rider*) upstream of the gene. However, the authors were unable to link the phenotype to the methylation and transcript data in the study.

In the following year, an independent epigenome-wide association study (EWAS) using a ~860,000 probe microarray was able to identify an epigenetic mark in a long interspersed nuclear element (LINE) retrotransposon related to the rice *Karma* in the intron of the *DEFICIENS* gene (Ong-Abdullah et al. 2015). The mark was identified using four sets of normal ramets (clonal palms), parthenocarpic mantled ramets and ortets from which the ramets were derived, from four independent tissue culture laboratories. Hypomethylation of the Karma element (*Bad Karma*) results in the production of a novel truncated kDEF1 peptide that terminates within the K domain of the DEFICIENS MADS box protein. The kDEF1 transcripts are only found in stage 3–5 of the developing parthenocarpic mantled inflorescence (Ong-Abdullah et al. 2015). Stage 3 is the transition stage between stage 2, where the floral anatomies are quite indistinguishable, and stage 4 where the elongation of the carpel is flanked by developmentally halted staminodes (rudimentary stamens) (Sarpan et al. 2015). The production of kDEF1 during this transition induces the mantled phenotype (Ong-Abdullah et al. 2015), which has carpel-like structures in place of staminodes at stage 4 (Sarpan et al. 2015). A cluster of antisense 24-nucleotide small interfering RNAs (siRNAs) was also reduced or absent in the Karma element of mantled palms (Ong-Abdullah et al. 2015). siRNAs can trigger DNA methylation at the homologous region in the genome using RNA-directed DNA methylation (Xie and Yu 2015). Good Karma (densely methylated *MANTLED* epialleles) can only be restored during shoot regeneration when there are still some elements of DNA methylation and small RNAs available during the tissue culture process (Ong-Abdullah et al. 2015). Moving forward, the development of the Karma diagnostic assay to predict the mantled phenotype will enable mantled palms to be culled at the nursery and therefore improve the confidence of the industry to produce clonal palms.

7.9 Bioinformatics Resources

Through the collaborative research under the Malaysian Oil Palm Genome Programme (MyOPGP), MPOB had sequenced and de novo assembled the two oil palm species, *E. guineensis* and *E. oleifera*. Oil palm transcriptomes from more than 30 samples were also sequenced (Singh et al. 2013b). These major breakthroughs have further enabled MPOB to discover commercially important genes (Singh et al. 2013a, 2014; Ong-Abdullah et al. 2015). The importance of these data makes it imperative that the data be collected in easily accessible Web portals and databases. Thus, to provide and disseminate high-quality updates of the oil palm genome information and data, a Web portal for oil palm genome resources (Genomsawit portal, http://genomsawit.mpob.gov.my) was developed (Fig. 7.9). Similar specialized crop or plant Web portals, such as the Date Palm Genomic Resource Database (DRDB) for the date palm genomics and molecular breeding (He et al. 2017), Gramene for cereal genomes sequencing programmes (Jaiswal 2006), Banana Genome Hub for Musa genomics (Droc et al. 2013), Brassica database (BRAD) for collection of various Brassica species genome information (Wang et al. 2015), Ensembl Plants (Kersey et al. 2012) and PlantGDB (Dong 2004) for various plant species, are also available.

The Genomsawit portal was designed as an initial access point for Web-based information systems, specialized databases and data analysis tools. The portal provides access to the two published oil palm genome sequences, *E. guineensis* and *E. oleifera*, and 30 transcriptome libraries, 22 of which were from *E. guineensis* and eight from *E. oleifera* (Singh et al. 2013b).

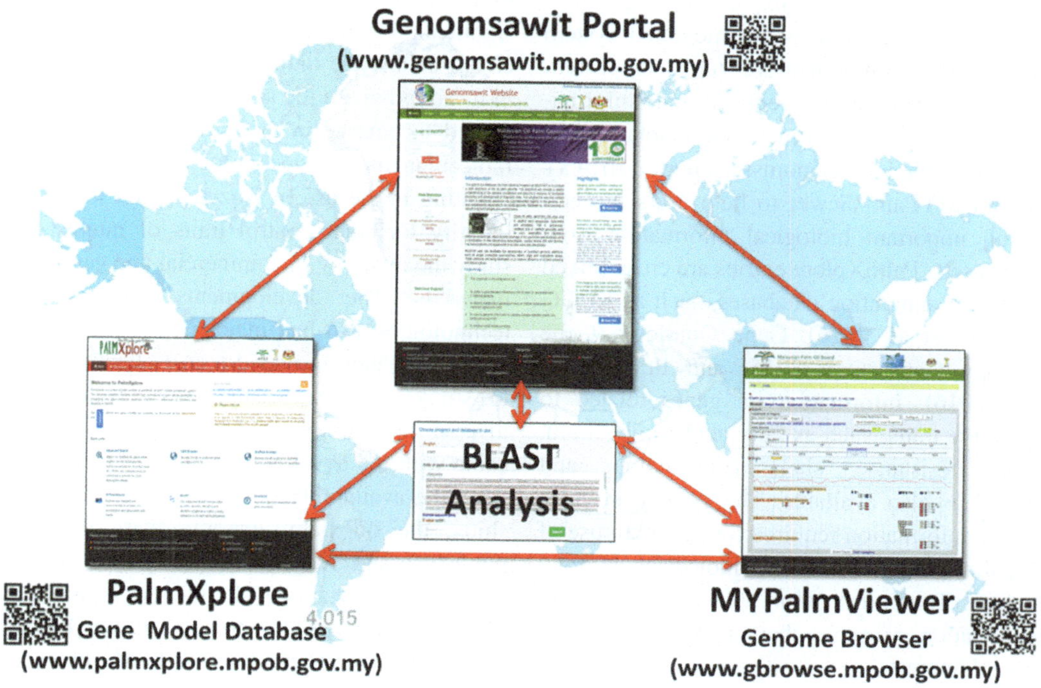

Fig. 7.9 Oil palm genome databases and Web portals

The P5 and EG5 genome builds are downloadable from the portal. As for the *E. oleifera* genome, two versions of the assemblies are also available: O7-build (1.374 Gb) and O8-build (1.402 Gb). Other oil palm genomics resources, such as gene sequence and annotations (Singh et al. 2013b; Chan et al. 2017b), transcripts (Singh et al. 2013b), hypomethylated genomic sequences (Low et al. 2014) and genomic markers (Ting et al. 2014, 2016), are also accessible.

The Web portal is also integrated with bioinformatics analysis tools, providing researchers with a platform to explore the genome data. NCBI BLAST and blastdbcmd are embedded into the portal to help researchers perform sequence comparison and sequence extraction of the oil palm genome data. Genomsawit provides 12 databases for BLAST analysis. The portal also contains a genome browser (MYPalmViewer; http://gbrowse.mpob.gov.my) based on the generic Genome Browser (GBrowse version 2.54) (Stein 2013) to visualize the annotation of EG5 and EG5.1 genome builds. There are 15 data tracks, such as comparative genomics, experimental results, GenBank sequences, gene predictions, markers, public database sequences and reference genomes, and features, such as retroelement, gaps, restriction site and general information available in MYPalmViewer. Researchers are able to perform search using keywords or location on the genome. MYPalmViewer also allows users to upload their annotation using GFF file format, and the tracks can be shared via email using the given URL of the tracks.

Improvements in high-throughput sequencing technologies had allowed the oil palm genome and transcripts to be sequenced, and subsequently used to generate predicted genes and its associated functional annotations. The association between genes and phenotypic traits is currently of great interest. A total of 26,059 high-confidence oil palm genes that were predicted using an integrated gene prediction pipeline (Chan et al. 2017a) was incorporated into the

PalmXplore (http://palmxplore.mpob.gov.my) database. The database is an integrated system that contains comprehensive tools to search, retrieve and browse the predicted gene information and its associated functional annotations. It is interoperable with Genomsawit portal, BLAST tool and MYPalmViewer to help users in deciphering important biological information from the data sets. PalmXplore entries are cross-linked to external annotation databases, such as KEGG (Kanehisa et al. 2016), Gene Ontology (Gene Ontology Consortium 2015) and Pfam (Finn et al. 2016). Information on genes related to imperative traits, such as fatty acid composition, disease resistance, shell, fruit colour and mantled, is also available. The system provides primary information required for optimum use of oil palm genome data for molecular biology research. The information will facilitate future identification of genes that are responsible for traits of interest.

The Genomsawit portal has been made available to the public since August 2013 and has become a one-stop centre for oil palm genome information. It provides sequence data and annotations from various resources, and bioinformatics tools to facilitate the use of the oil palm genome data. The portal will be regularly updated with the latest genome data, setting the stage for further genetic and breeding studies.

7.10 Conclusion

Genetic improvement of oil palm through conventional breeding is extremely slow and costly, as the breeding cycle can take up to 12 years. Due to this, the oil palm research community and industry members started looking at different technologies to improve yield. Although the first major breakthrough was the discovery of the monogenetic inheritance of the *SHELL* gene, the first biotechnology breakthrough only happened in the 1970s with the discovery of vegetative propagation of oil palm. Nevertheless, its implementation slowed down when somaclonal abnormalities were identified. Although the tissue culture process has since been improved to significantly reduce the occurrence of abnormal clonal palms, the industry is now more cautious in its adoption. At the same time, new techniques started to emerge in the early 1980s that made molecular marker technologies more feasible, resulting in the use of isozymes and markers to identify loci that control traits of interest, and study the diversity of commercial and germplasm materials. The developments of sequencing technologies also had major impact on oil palm research as they provided a cost-effective means to study gene expression and develop more diverse sets of markers. The major breakthrough came with the development of NGS technology. With the availability of NGS and smart partnerships, the oil palm genome was sequenced, published and released to the public domain. This is an important development as it provided a platform for more comprehensive and collaborative research programmes to be developed. It also forms an important resource for the global research community. As resources and human capital are limited, more collaborative research will maximize the outcome of projects. Collaboration is the key to unlocking the potential of not only the data but the capabilities of the researchers involved. It will also pave the way for the development of more tools and services that can contribute to the oil palm industry. One good example is the collaborative programme that led to the discoveries of *SHELL*, fruit colour and *MANTLED* genes. These discoveries have since been translated to practical applications in the field. The SHELL diagnostic assay can be used to screen out contaminants in commercial planting materials as early as in the seed and nursery stages, and streamline breeding programmes, while VIR diagnostic assays can be used to develop new homozygous planting materials that have future implications on labour, harvesting standards and the quality of fruits harvested. Karma diagnostic assays will provide the industry with the confidence to increase the production of clonal palms, which have the potential to increase yields by 20–30%. All these discoveries have far-reaching impact on the continuous improvement of breeding materials and the continued sustainable growth of the oil palm industry.

References

Aberlenc-Bertossi F, Chabrillange N, Duval Y, Tregear J (2008) Contrasting globulin and cysteine proteinase gene expression patterns reveal fundamental developmental differences between zygotic and somatic embryos of oil palm. Tree Physiol 28(8):1157–1167. https://doi.org/10.1093/treephys/28.8.1157

Abrizah O, Lazarus C, Fraser T, Stobart K (2000) Cloning of a palmitoyl-acyl carrier protein thioesterase from oil palm. Biochem Soc Trans 28(6):619–622. https://doi.org/10.1042/BST0280619

Adam H, Jouannic S, Escoute J, Duval Y, Verdeil JL, Tregear JW (2005) Reproductive developmental complexity in the African oil palm (*Elaeis guineensis*, Arecaceae). Am J Bot 92(11):1836–1852

Adam H, Jouannic S, Morcillo F, Richaud F, Duval Y, Tregear JW (2006) MADS box genes in oil palm (*Elaeis guineensis*): Patterns in the evolution of the *SQUAMOSA*, *DEFICIENS*, *GLOBOSA*, *AGAMOUS*, and *SEPALLATA* subfamilies. J Mol Evol 62(1):15–31. https://doi.org/10.1007/s00239-005-0333-7

Adam H, Jouannic S, Orieux Y, Morcillo F, Richaud F, Duval Y, Tregear JW (2007) Functional characterization of MADS box genes involved in the determination of oil palm flower structure. J Exp Bot 58 (6):1245–1259. https://doi.org/10.1093/jxb/erl263

Adams MD, Kelley JM, Gocayne JD, Dubnick M, Polymeropoulos MH, Xiao H, Merril CR, Wu A, Olde B, Moreno RF, Et A (1991) Complementary DNA sequencing: expressed sequence tags and human genome project. Science 252(5013):1651–1656. https://doi.org/10.1126/science.2047873

Alban C, Job D, Douce R (2000) Biotin metabolism in plants. Annu Rev Plant Physiol Plant Mol Biol 51 (1):17–47. https://doi.org/10.1146/annurev.arplant.51.1.17

Alexandrov NN, Brover VV, Freidin S, Troukhan ME, Tatarinova TV, Zhang H, Swaller TJ, Lu Y-PP, Bouck J, Flavell RB, Feldmann KA (2009) Insights into corn genes derived from large-scale cDNA sequencing. Plant Mol Biol 69(1–2):179–194. https://doi.org/10.1007/s11103-008-9415-4

Al-Mssallem IS, Hu S, Zhang X, Lin Q, Liu W, Tan J, Yu X, Liu J, Pan L, Zhang T, Yin Y, Xin C, Wu H, Zhang G, Ba Abdullah MM, Huang D, Fang Y, Alnakhli YO, Jia S, Yin A, Alhuzimi EM, Alsaihati BA, Al-Owayyed SA, Zhao D, Zhang S, Al-Otaibi NA, Sun G, Majrashi MA, Li F, Wang J, Yun Q, Alnassar NA, Wang L, Yang M, Al-Jelaify RF, Liu K, Gao S, Chen K, Alkhaldi SR, Liu G, Zhang M, Guo H, Yu J (2013) Genome sequence of the date palm *Phoenix dactylifera* L. Nat Commun 4:2274. https://doi.org/10.1038/ncomms3274

Alwee SS, Van Der Linden CG, Van Der Schoot J, De Folter S, Angenent GC, Cheah SC, Smulders MJM (2006) Characterization of oil palm MADS box genes in relation to the mantled flower abnormality. Plant Cell, Tissue Organ Cult 85(3):331–344. https://doi.org/10.1007/s11240-006-9084-4

Bai B, Wang L, Lee M, Zhang Y, Alfiko Y, Ye BQ, Wan ZY, Lim CH, Suwanto A, Chua NH, Yue GH (2017) Genome-wide identification of markers for selecting higher oil content in oil palm. BMC Plant Biol 17:93. https://doi.org/10.1186/s12870-017-1045-Z

Bai B, Wang L, Zhang YJ, Lee M, Rahmadsyah R, Alfiko Y, Ye BQ, Purwantomo S, Suwanto A, Chua NH, Yue GH (2018) Developing genome-wide SNPs and constructing an ultrahigh-density linkage map in oil palm. Sci Rep 8:691. https://doi.org/10.1038/s41598-017-18613-2

Beirnaert A, Vanderweyen R (1941) Contribution a` l'e´tude ge´ne´tique et biome´trique des varie´tie´s d'Elaeis guineensis Jacq. Publ Inst Nat Etude Agron Congo Belge Ser Sci 27:1–101

Belkhadir Y, Subramaniam R, Dangl JL (2004) Plant disease resistance protein signaling: NBS–LRR proteins and their partners. Curr Opin Plant Biol 7(4):391–399. https://doi.org/10.1016/j.pbi.2004.05.009

Beulé T, Camps C, Debiesse S, Tranchant C, Dussert S, Sabau X, Jaligot E, Alwee SSRS, Tregear JW (2011) Transcriptome analysis reveals differentially expressed genes associated with the mantled homeotic flowering abnormality in oil palm (*Elaeis guineensis*). Tree Genet Genomes 7(1):169–182. https://doi.org/10.1007/s11295-010-0323-9

Beulé T, Agbessi MDT, Dussert S, Jaligot E, Guyot R (2015) Genome-wide analysis of LTR-retrotransposons in oil palm. BMC Genom 16:795. https://doi.org/10.1186/s12864-015-2023-1

Billotte N, Risterucci AM, Barcelos E, Noyer JL, Amblard P, Baurens FC (2001) Development, characterisation, and across-taxa utility of oil palm (*Elaeis guineensis* Jacq.) microsatellite markers. Genome 44 (3):413–425. https://doi.org/10.1139/gen-44-3-413

Billotte N, Marseillac N, Risterucci A-M, Adon B, Brottier P, Baurens F-C, Singh R, Herrán A, Asmady H, Billot C, Amblard P, Durand-Gasselin T, Courtois B, Asmono D, Cheah SC, Rohde W, Ritter E, Charrier A (2005) Microsatellite-based high density linkage map in oil palm (*Elaeis guineensis* Jacq.). Theor Appl Genet 110(4):754–765. https://doi.org/10.1007/s00122-004-1901-8

Billotte N, Jourjon MF, Marseillac N, Berger A, Flori A, Asmady H, Adon B, Singh R, Nouy B, Potier F, Cheah SC, Rohde W, Ritter E, Courtois B, Charrier A, Mangin B (2010) QTL detection by multi-parent linkage mapping in oil palm (*Elaeis guineensis* Jacq.). Theor Appl Genet 120(8):1673–1687. https://doi.org/10.1007/s00122-010-1284-y

Bonaldo MF, Lennon G, Soares MB (1996) Normalization and subtraction: two approaches to facilitate gene discovery. Genome Res 6(9):791–806. https://doi.org/10.1101/gr.6.9.791

Borevitz JO, Xia Y, Blount J, Dixon RA, Lamb C (2000) Activation tagging identifies a conserved MYB regulator of phenylpropanoid biosynthesis. Plant Cell 12 (12):2383–2394. https://doi.org/10.2307/3871236

Bourgis F, Kilaru A, Cao X, Ngando-Ebongue G-F, Drira N, Ohlrogge JB, Arondel V (2011) Comparative transcriptome and metabolite analysis of oil palm and date palm mesocarp that differ dramatically in carbon partitioning. Proc Natl Acad Sci 108(30):12527–12532. https://doi.org/10.1073/pnas.1106502108

Budiman MA, Singh R, Low ETL, Nunberg A, Citek R, Rohlfing TJ, Bedell A, Lakey ND, Martienssen RA, Cheah CS (2005) Sequencing of the oil palm genespace. In: Proceedings of PIPOC2005 international palm oil congress (agriculture, biotechnology and sustainability conference). Petaling Jaya, Selangor, Malaysia, pp 628–639

Campbell WH, Gowri G (1990) Codon usage in higher plants, green algae, and cyanobacteria. Plant Physiol 92(1):1–11. https://doi.org/10.1104/pp.92.1.1

Castilho AM, Vershinin AV, Heslop-Harrison JS (2000) Repetitive DNA and the chromosomes in the genome of oil palm (*Elaeis guineensis*). Ann Bot 85(6):837–844. https://doi.org/10.1006/anbo.2000.1145

Chan PL, Ma LS, Low ETL, Shariff EM, Ooi LCL, Cheah SC, Singh R (2010) Normalized embryoid cDNA library of oil palm (*Elaeis guineensis*). Electron J Biotechnol 13(1):14. https://doi.org/10.2225/vol13-issue1-fulltext-14

Chan K-L, Rosli R, Tatarinova TV, Hogan M, Firdaus-Raih M, Low E-TL (2017a) Seqping: Gene prediction pipeline for plant genomes using self-training gene models and transcriptomic data. BMC Bioinformatics 18(Suppl1):29. https://doi.org/10.1186/s12859-016-1426-6

Chan K-L, Tatarinova TV, Rosli R, Amiruddin N, Azizi N, Halim MAA, Sanusi NSNM, Jayanthi N, Ponomarenko P, Triska M, Solovyev V, Firdaus-Raih M, Sambanthamurthi R, Murphy D, Low E-TL (2017b) Evidence-based gene models for structural and functional annotations of the oil palm genome. Biol Direct 12(1):21. https://doi.org/10.1186/s13062-017-0191-4

Chin CW (1995) Oil palm planting materials and quality control. In: Proceedings of the 1995 PORIM national oil conference-tech in plant, pp 38–47

Clément Y, Fustier M-A, Nabholz B, Glémin S (2015) The bimodal distribution of genic GC content is ancestral to monocot species. Genome Biol Evol 7 (1):336–348. https://doi.org/10.1093/gbe/evu278

Cochard B, Durand-Gasselin T, Amblard P, Konan EK, Gogor S (1999) Performance of adult oil palm clones. In: Ariffin D, Chan KW, Sharifah SSA (eds) Emerging technologies and opportunities in the next millennium. Agriculture conference: proceedings of 1999 PORIM International Palm Oil Congress. PORIM, Kuala Lumpur, pp 53–64

Corley RHV (2005) Illegitimacy in oil palm breeding—a review. J Oil Palm Res 17:64–69

Corley RHV, Law IH (1997) The future for oil palm clones. In: Pushparajah E (ed) Plantation Management for the 21st Century. Incorporated Society of Planters, pp 279–289

Corley RHV, Lee CH, Law IM, Wong CY (1986) Abnormal flower development in oil palm clones. Planter 62(723):233–240

D'Hont A, Denoeud F, Aury J-M, Baurens F-C, Carreel F, Garsmeur O, Noel B, Bocs S, Droc G, Rouard M, Da Silva C, Jabbari K, Cardi C, Poulain J, Souquet M, Labadie K, Jourda C, Lengellé J, Rodier-Goud M, Alberti A, Bernard M, Correa M, Ayyampalayam S, Mckain MR, Leebens-Mack J, Burgess D, Freeling M, Mbéguié-A-Mbéguié D, Chabannes M, Wicker T, Panaud O, Barbosa J, Hribova E, Heslop-Harrison P, Habas R, Rivallan R, Francois P, Poiron C, Kilian A, Burthia D, Jenny C, Bakry F, Brown S, Guignon V, Kema G, Dita M, Waalwijk C, Joseph S, Dievart A, Jaillon O, Leclercq J, Argout X, Lyons E, Almeida A, Jeridi M, Dolezel J, Roux N, Risterucci A-M, Weissenbach J, Ruiz M, Glaszmann J-C, Quétier F, Yahiaoui N, Wincker P (2012) The banana (*Musa acuminata*) genome and the evolution of monocotyledonous plants. Nature 488(7410):213–217. https://doi.org/10.1038/nature11241

Devuyst A (1953) Selection of the oil palm (*Elaeis guineensis*) in Africa. Nature 172:685–686. https://doi.org/10.1038/172685a0

Dong Q (2004) PlantGDB, plant genome database and analysis tools. Nucleic Acids Res 32(90001):354D–459D. https://doi.org/10.1093/nar/gkh046

Droc G, Larivière D, Guignon V, Yahiaoui N, This D, Garsmeur O, Dereeper A, Hamelin C, Argout X, Dufayard J-F, Lengelle J, Baurens F-C, Cenci A, Pitollat B, D'Hont A, Ruiz M, Rouard M, Bocs S (2013) The Banana Genome Hub. Database 2013:bat035. https://doi.org/10.1093/database/bat035

Durand-Gasselin T, Le Guen VL, Konan E, Duval Y (1990) Oil palm (*Elaeis guineensis* Jacq.) plantations in Cote d'Ivoire obtained through in vitro culture—First results. Ole´agineux 45:1–11

Dussert S, Guerin C, Andersson M, Joët T, Tranbarger TJ, Pizot M, Sarah G, Omore A, Durand-Gasselin T, Morcillo F (2013) Comparative transcriptome analysis of three oil palm fruit and seed tissues that differ in oil content and fatty acid composition. Plant Physiol 162 (3):1337–1358. https://doi.org/10.1104/pp.113.220525

Eilbeck K, Moore B, Holt C, Yandell M (2009) Quantitative measures for the management and comparison of annotated genomes. BMC Bioinformatics 10(1):67. https://doi.org/10.1186/1471-2105-10-67

Elhaik E, Pellegrini M, Tatarinova TV (2014) Gene expression and nucleotide composition are associated with genic methylation level in *Oryza sativa*. BMC

Bioinform 15(1):23. https://doi.org/10.1186/1471-2105-15-23

Fawcett JA, Maere S, Van de Peer Y (2009) Plants with double genomes might have had a better chance to survive the Cretaceous-Tertiary extinction event. Proc Natl Acad Sci 106(14):5737–5742. https://doi.org/10.1073/pnas.0900906106

Filho JAF, de Brito LS, Leão AP, Alves AA, Formighieri EF, Souza MT (2017) In silico approach for characterization and comparison of repeats in the genomes of oil and date palms. Bioinform Biol Insights 11:117793221770238. https://doi.org/10.1177/1177932217702388

Finn RD, Coggill P, Eberhardt RY, Eddy SR, Mistry J, Mitchell AL, Potter SC, Punta M, Qureshi M, Sangrador-Vegas A, Salazar GA, Tate J, Bateman A (2016) The Pfam protein families database: towards a more sustainable future. Nucleic Acids Res 44(D1): D279–D285. https://doi.org/10.1093/nar/gkv1344

Gaj MD, Zhang S, Harada JJ, Lemaux PG (2005) Leafy cotyledon genes are essential for induction of somatic embryogenesis of *Arabidopsis*. Planta 222(6):977–988. https://doi.org/10.1007/s00425-005-0041-y

Galperin MY, Cochrane GR (2009) Nucleic acids research annual database issue and the NAR online molecular biology database collection in 2009. Nucleic Acids Res 37:D1–D4. https://doi.org/10.1093/nar/gkn942

Gan ST, Wong WC, Wong CK, Soh AC, Kilian A, Low E-TL, Massawe F, Mayes S (2018) High density SNP and DArT-based genetic linkage maps of two closely related oil palm populations. J Appl Genet 59(1):23–34. https://doi.org/10.1007/s13353-017-0420-7

Gene Ontology Consortium (2015) Gene ontology consortium: going forward. Nucleic Acids Res 43(D1): D1049–D1056. https://doi.org/10.1093/nar/gku1179

Gonzalez A, Zhao M, Leavitt JM, Lloyd AM (2008) Regulation of the anthocyanin biosynthetic pathway by the TTG1/bHLH/Myb transcriptional complex in *Arabidopsis* seedlings. Plant J 53:814–827. https://doi.org/10.1111/j.1365-313X.2007.03373.x

Hartley CWS (1988) The oil palm (*Elaeis guineesis* Jacq.). Longman Scientific & Technical, England

He Z, Zhang C, Liu W, Lin Q, Wei T, Aljohi HA, Chen W-H, Hu S (2017) DRDB: an online date palm genomic resource database. Front Plant Sci 8:1889. https://doi.org/10.3389/fpls.2017.01889

Hershberg R, Petrov DA (2008) Selection on codon bias. Annu Rev Genet 42(1):287–299. https://doi.org/10.1146/annurev.genet.42.110807.091442

Ho CL, Kwan YY, Choi MC, Tee S-SS, Ng W-HH, Lim KA, Lee Y-PP, Ooi SE, Lee W-WW, Tee JM, Tan SH, Kulaveerasingam H, Alwee SSRS, Abdullah MO (2007) Analysis and functional annotation of expressed sequence tags (ESTs) from multiple tissues of oil palm (*Elaeis guineensis* Jacq.). BMC Genomics 8:381. doi: https://doi.org/10.1186/1471-2164-8-381

Ho C-L, Tan Y-C, Yeoh K-A, Ghazali A-K, Yee W-Y, Hoh C-C (2016) *De novo* transcriptome analyses of host-fungal interactions in oil palm (*Elaeis guineensis* Jacq.). BMC Genomics 17(1):66. https://doi.org/10.1186/s12864-016-2368-0

Ho H, Gudimella R, Ong-Abdullah M, Harikrishna JA (2017) Expression of microRNAs during female inflorescence development in African oil palm (*Elaeis guineensis* Jacq.). Tree Genet Genomes 13(2):35. https://doi.org/10.1007/s11295-017-1120-5

Ho C-L, Tan Y-C, Yeoh K-A, Lee W-K, Ghazali A-K, Yee W-Y, Hoh C-C (2018) Transcriptional response of oil palm (*Elaeis guineensis* Jacq.) inoculated simultaneously with both *Ganoderma boninense* and *Trichoderma harzianum*. Plant Gene 13:56–63. https://doi.org/10.1016/j.plgene.2018.01.003

Holt C, Yandell M (2011) MAKER2: an annotation pipeline and genome-database management tool for second-generation genome projects. BMC Bioinform 12(1):491. https://doi.org/10.1186/1471-2105-12-491

Ikeda M, Umehara M, Kamada H (2006) Embryogenesis-related genes; Its expression and roles during somatic and zygotic embryogenesis in carrot and *Arabidopsis*. Plant Biotechnol 23(2):153–161. https://doi.org/10.5511/plantbiotechnology.23.153

Iriarte A, Baraibar JD, Romero H, Castro-Sowinski S, Musto H (2013) Evolution of optimal codon choices in the family Enterobacteriaceae. Microbiology 159 (Pt_3):555–564. https://doi.org/10.1099/mic.0.061952-0

Jaiswal P (2006) Gramene: a bird's eye view of cereal genomes. Nucleic Acids Res 34(90001):D717–D723. https://doi.org/10.1093/nar/gkj154

Jaligot E, Rival A, Beulé T, Dussert S, Verdeil J-L (2000) Somaclonal variation in oil palm (*Elaeis guineensis* Jacq.): the DNA methylation hypothesis. Plant Cell Rep 19(7):684–690. https://doi.org/10.1007/s002999900177

Jaligot E, Hooi WY, Debladis E, Richaud F, Beulé T, Collin M, Agbessi MDT, Sabot F, Garsmeur O, D'Hont A, Alwee SSRS, Rival A (2014) DNA methylation and expression of the EgDEF1 gene and neighboring retrotransposons in mantled somaclonal variants of oil palm. PLoS ONE 9(3):e91896. https://doi.org/10.1371/journal.pone.0091896

Jiao Y, Li J, Tang H, Paterson AH (2014) Integrated syntenic and phylogenomic analyses reveal an ancient genome duplication in monocots. Plant Cell 26 (7):2792–2802. https://doi.org/10.1105/tpc.114.127597

Jin J, Lee M, Bai B, Sun Y, Qu J, Rahmadsyah Alfiko Y, Lim CH, Suwanto A, Sugiharti M, Wong L, Ye J, Chua N-H, Yue GH (2016) Draft genome sequence of an elite Dura palm and whole-genome patterns of DNA variation in oil palm. DNA Res 23(6):527–533. https://doi.org/10.1093/dnares/dsw036

Jin J, Sun Y, Qu J, Syah R, Lim C-H, Alfiko Y, Rahman N, Suwanto A, Yue G, Wong L, Chua N-H, Ye J (2017) Transcriptome and functional analysis reveals hybrid vigor for oil biosynthesis in oil palm. Sci Rep 7(1):439. https://doi.org/10.1038/s41598-017-00438-8

Jones L (1974) Propagation of clonal palms by tissue culture. Oil Palm News 17:1–8

Jouannic S, Argout X, Lechauve F, Fizames C, Borgel A, Morcillo F, Aberlenc-Bertossi F, Duval Y, Tregear J (2005) Analysis of expressed sequence tags from oil palm (*Elaeis guineensis*). FEBS Lett 579(12):2709–2714. https://doi.org/10.1016/j.febslet.2005.03.093

Jurka J, Kapitonov VV, Pavlicek A, Klonowski P, Kohany O, Walichiewicz J (2005) Repbase update, a database of eukaryotic repetitive elements. Cytogenet Genome Res 110(1–4):462–467. https://doi.org/10.1159/000084979

Kanchanapoom K, Domyoas P (1999) The origin and development of embryoids in oil palm (*Elaeis guineensis* Jacq) embryo culture. Science Asia 25 (4):195. https://doi.org/10.2306/scienceasia1513-1874.1999.25.195

Kanehisa M, Sato Y, Kawashima M, Furumichi M, Tanabe M (2016) KEGG as a reference resource for gene and protein annotation. Nucleic Acids Res 44 (D1):D457–D462. https://doi.org/10.1093/nar/gkv1070

Kersey PJ, Staines DM, Lawson D, Kulesha E, Derwent P, Humphrey JC, Hughes DST, Keenan S, Kerhornou A, Koscielny G, Langridge N, McDowall MD, Megy K, Maheswari U, Nuhn M, Paulini M, Pedro H, Toneva I, Wilson D, Yates A, Birney E (2012) Ensembl Genomes: an integrative resource for genome-scale data from non-vertebrate species. Nucleic Acids Res 40(D1):D91–D97. https://doi.org/10.1093/nar/gkr895

Khaw CH, Ng SK (1998) Performance of commercial scale clonal oil palm (*Elaeis guineensis* Jacq.) plantings in Malaysia. Acta Hortic 461:251–258. https://doi.org/10.17660/ActaHortic.1998.461.27

Korf I (2004) Gene finding in novel genomes. BMC Bioinformatics 5:59. https://doi.org/10.1186/1471-2105-5-59

Kubis SE, Castilho AMMF, Vershinin AV, Heslop-Harrison JSP (2003) Retroelements, transposons and methylation status in the genome of oil palm (*Elaeis guineensis*) and the relationship to somaclonal variation. Plant Mol Biol 52(1):69–79. https://doi.org/10.1023/A:1023942309092

Kushairi A, Rajanaidu N (2000) Breeding populations, seed production and nursery management. In: Basiron Y, Jalani BS, Chan KW (eds) Advances in oil palm research, pp 39–98

Kwong QB, Teh CK, Ong AL, Chew FT, Mayes S, Kulaveerasingam H, Tammi M, Yeoh SH, Appleton DR, Harikrishna JA (2017) Evaluation of methods and marker Systems in Genomic Selection of oil palm (*Elaeis guineensis* Jacq.). BMC Genet 18(1):107. https://doi.org/10.1186/s12863-017-0576-5

Latiff A (2000) The biology of the genus *Elaeis*. In: Basiron Y, Jalani BS, Chan KW (eds) Advances in oil palm research, vol 1. Malaysian Palm Oil Board, Selangor, pp 19–38

Lee M, Xia JH, Zou Z, Ye J, Rahmadsyah Alfiko Y, Jin J, Lieando JV, Purnamasari MI, Lim CH, Suwanto A, Wong L, Chua N-H, Yue GH (2015) A consensus linkage map of oil palm and a major QTL for stem height. Sci Rep 5(1):8232. https://doi.org/10.1038/srep08232

Li R, Reddy VA, Jin J, Rajan C, Wang Q, Yue G, Lim CH, Chua N-H, Ye J, Sarojam R (2017) Comparative transcriptome analysis of oil palm flowers reveals an EAR-motif-containing R2R3-MYB that modulates phenylpropene biosynthesis. BMC Plant Biol 17(1):219. https://doi.org/10.1186/s12870-017-1174-4

Lin H-C, Morcillo F, Dussert S, Tranchant-Dubreuil C, Tregear JW, Tranbarger TJ (2009) Transcriptome analysis during somatic embryogenesis of the tropical monocot *Elaeis guineensis*: evidence for conserved gene functions in early development. Plant Mol Biol 70(1–2):173–192. https://doi.org/10.1007/s11103-009-9464-3

Llorens C, Futami R, Covelli L, Domínguez-Escribá L, Viu JM, Tamarit D, Aguilar-Rodríguez J, Vicente-Ripolles M, Fuster G, Bernet GP, Maumus F, Munoz-Pomer A, Sempere JM, Latorre A, Moya A (2011) The gypsy database (GyDB) of mobile genetic elements: release 2.0. Nucleic Acids Res 39:D70–D74. https://doi.org/10.1093/nar/gkq1061

Low ETL, Tan JS, Chan PL, Boon SH, Wong YL, Rozana R, Ooi L-L, Ma LS, Ong-Abdullah M, Cheah SC, Rajinder S (2006) Developments toward the application of DNA chip technology in oil palm tissue culture. J Oil Palm Res (Special Issue-April 2006):87–98

Low E-TL, Alias H, Boon S-H, Shariff EM, Tan C-YA, Ooi LC, Cheah S-C, Raha A-R, Wan K-L, Singh R (2008) Oil palm (*Elaeis guineensis* Jacq.) tissue culture ESTs: Identifying genes associated with callogenesis and embryogenesis. BMC Plant Biol 8 (1):62. https://doi.org/10.1186/1471-2229-8-62

Low E-TL, Rosli R, Jayanthi N, Mohd-Amin AH, Azizi N, Chan K-L, Maqbool NJ, Maclean P, Brauning R, McCulloch A, Moraga R, Ong-Abdullah M, Singh R (2014) Analyses of hypomethylated oil palm gene space. PLoS ONE 9(1):e86728. https://doi.org/10.1371/journal.pone.0086728

Low ETL, Singh R, Rajanaidu N, Ong-Abdullah M, Ooi LCL, Lakey ND, Smith SW, Ordway JMJ, Sambanthamurthi R (2016) New frontiers for the oil palm industry through genome technology. Planter 92 (1087):701–710

Low E-TL, Jayanthi N, Chan K-L, Sanusi NSNM, Halim MAA, Rosli R, Azizi N, Amiruddin N, Angel LPL, Ong-Abdullah M, Singh R, Manaf MAA, Sambanthamurthi R, Parveez GKA, Kushairi A (2017) The oil palm genome revolution. J Oil Palm Res 29(4):456–468. https://doi.org/10.21894/jopr.2017.00018

Maizura I, Rajanaidu N, Zakri AH, Cheah SC (2006) Assessment of genetic diversity in oil palm (*Elaeis guineensis* Jacq.) using Restriction Fragment Length Polymorphism (RFLP). Genet Resour Crop Evol 53 (1):187–195. https://doi.org/10.1007/s10722-004-4004-0

Majoros WH, Pertea M, Salzberg SL (2004) TigrScan and GlimmerHMM: Two open source *ab initio* eukaryotic gene-finders. Bioinformatics 20(16):2878–2879. https://doi.org/10.1093/bioinformatics/bth315

Malaysian Palm Oil Board (2017) Review of the Malaysian oil palm industry 2016

Mathew LS, Spannagl M, Al-Malki A, George B, Torres MF, Al-Dous EK, Al-Azwani EK, Hussein E, Mathew S, Mayer KF, Mohamoud Y, Suhre K, Malek JA (2014) A first genetic map of date palm (*Phoenix dactylifera*) reveals long-range genome structure conservation in the palms. BMC Genom 15 (1):285. https://doi.org/10.1186/1471-2164-15-285

Matthes M, Singh R, Cheah S-C, Karp A (2001) Variation in oil palm (*Elaeis guineensis* Jacq.) tissue culture-derived regenerants revealed by AFLPs with methylation-sensitive enzymes. TAG Theor Appl Genet 102(6–7):971–979. https://doi.org/10.1007/s001220000491

Mayes S, Jack PL, Corley RHV, Marshall DF (1997) Construction of a RFLP genetic linkage map for oil palm (*Elaeis guineensis* Jacq.). Genome 40(1):116–122. https://doi.org/10.1139/g97-016

Mazumdar P, Binti Othman R, Mebus K, Ramakrishnan N, Ann Harikrishna J (2017) Codon usage and codon pair patterns in non-grass monocot genomes. Ann Bot 120(6):893–909. https://doi.org/10.1093/aob/mcx112

McClintock B (1951) Chromosome organization and genic expression. Cold Spring Harb Symp Quant Biol 16:13–47. https://doi.org/10.1101/SQB.1951.016.01.004

McHale L, Tan X, Koehl P, Michelmore RW (2006) Plant NBS-LRR proteins: adaptable guards. Genome Biol 7 (4):212. https://doi.org/10.1186/gb-2006-7-4-212

McKain MR, Tang H, McNeal JR, Ayyampalayam S, Davis JI, DePamphilis CW, Givnish TJ, Pires JC, Stevenson DW, Leebens-Mack JH (2016) A phylogenomic assessment of ancient polyploidy and genome evolution across the Poales. Genome Biol Evol 8 (4):1150–1164. https://doi.org/10.1093/gbe/evw060

Montoya C, Lopes R, Flori A, Cros D, Cuellar T, Summo M, Espeout S, Rivallan R, Risterucci A-M, Bittencourt D, Zambrano JR, Alarcón G WH, Villeneuve P, Pina M, Nouy B, Amblard P, Ritter E, Leroy T, Billotte N (2013) Quantitative trait loci (QTLs) analysis of palm oil fatty acid composition in an interspecific pseudo-backcross from *Elaeis oleifera* (H.B.K.) Cortés and oil palm (*Elaeis guineensis* Jacq.). Tree Genet Genomes 9(5):1207–1225. https://doi.org/10.1007/s11295-013-0629-5

Morcillo F, Hartmann C, Duval Y, Tregear JW (2001) Regulation of 7S globulin gene expression in zygotic and somatic embryos of oil palm. Physiol Plant 112 (2):233–243. https://doi.org/10.1034/j.1399-3054.2001.1120212.x

Morcillo F, Gallard A, Pillot M, Jouannic S, Aberlenc-Bertossi F, Collin M, Verdeil JL, Tregear JW (2007) EgAP2-1, an AINTEGUMENTA-like (AIL) gene expressed in meristematic and proliferating tissues of embryos in oil palm. Planta 226(6):1353–1362. https://doi.org/10.1007/s00425-007-0574-3

Moretzsohn MC, Nunes CDM, Ferreira ME, Grattapaglia D (2000) RAPD linkage mapping of the shell thickness locus in oil palm (*Elaeis guineensis* Jacq.). TAG Theor Appl Genet 100(1):63–70. https://doi.org/10.1007/s001220050009

NCBI (2018) txid51952[Organism:exp]—EST—NCBI. https://www.ncbi.nlm.nih.gov/nucest/?term=txid51952[Organism:exp]. Accessed 9 Jul 2018

Nurniwalis AW, Suhaimi N, Siti Nor Akmar A, Aminah S, Mohamad Arif MA (2008) Gene discovery via expressed sequence tags from the oil palm (*Elaeis guineensis* Jacq.) mesocarp. J Oil Palm Res 87–96

Okoye MN, Bakoumé C, Uguru MI, Singh R, Okwuagwu CO (2016) Genetic relationships between elite oil palms from Nigeria and selected breeding and germplasm materials from Malaysia via simple sequence repeat (SSR) markers. J Agric Sci 8 (2):159. https://doi.org/10.5539/jas.v8n2p159

Omar WSW, Willis LB, Rha C, Sinskey AJ, Ramli US, Yunus AMM, Parveez GKA, Sambanthamurthi R (2008) Isolation and utilization of acetyl-coa carboxylase from oil palm (*Elaeis guineensis*) mesocarp. J Oil Palm Res 2:97–107

Ong PW, Maizura I, Abdullah NAP, Rafii MY, Ooi LCL, Low ETL, Singh R (2015) Development of SNP markers and their application for genetic diversity analysis in the oil palm (*Elaeis guineensis*). Genet Mol Res 14(4):12205–12216. https://doi.org/10.4238/2015.October.9.9

Ong PW, Maizura I, Marhalil M, Rajanaidu N, Abdullah NAP, Rafii MY, Ooi LCL, Low ETL, Singh R (2018) Association of SNP markers with height increment in mpob-angolan natural oil palm populations. J Oil Palm Res 30:61–70. https://doi.org/10.21894/jopr.2017.0003

Ong-Abdullah M, Ooi S (2006) Biomarkers: finding a niche in oil palm tissue culture. Part 1—laying the foundation. Oil Palm Bull 53:36–48

Ong-Abdullah M, Ordway JM, Jiang N, Ooi S-E, Kok S-Y, Sarpan N, Azimi N, Hashim AT, Ishak Z, Rosli SK, Malike FA, Bakar NAA, Marjuni M, Abdullah N, Yaakub Z, Amiruddin MD, Nookiah R, Singh R, Low E-TL, Chan K-L, Azizi N, Smith SW, Bacher B, Budiman MA, Van Brunt A, Wischmeyer C, Beil M, Hogan M, Lakey N, Lim C-C, Arulandoo X, Wong C-K, Choo C-N, Wong W-C, Kwan Y-Y, Alwee SSRS, Sambanthamurthi R, Martienssen RA (2015) Loss of Karma transposon methylation underlies the mantled somaclonal variant of oil palm. Nature 525(7570):533–537. https://doi.org/10.1038/nature15365

Ong-Abdullah M, Ordway JM, Nan J, Ooi SE, Mokri A, Kok SY, Sarpan N, Azimi N, Hashim AT, Ishak Z, Rosli SK, Nookiah R, Singh R, Low ETL, Sachdeva M, Smith SW, Lakey N, Martienssen RA, Sambanthamurthi R (2016) Tissue culture and epigenetics. Planter 92(1087):741–749

Ooi LC-L, Low E-TL, Abdullah MO, Nookiah R, Ting NC, Nagappan J, Manaf MAA, Chan K-L,

Halim MA, Azizi N, Omar W, Murad AJ, Lakey N, Ordway JM, Favello A, Budiman MA, Van Brunt A, Beil M, Leininger MT, Jiang N, Smith SW, Brown CR, Kuek ACS, Bahrain S, Hoynes-O'Connor A, Nguyen AY, Chaudhari HG, Shah SA, Choo Y-M, Sambanthamurthi R, Singh R (2016) Non-tenera contamination and the economic impact of SHELL genetic testing in the Malaysian independent oil palm industry. Front Plant Sci 7:771. https://doi.org/10.3389/fpls.2016.00771

Osborne DJ, Henderson J, Corley RH (1992) Controlling fruit-shedding in the oil palm. Endeavour 16(4):173–177. https://doi.org/10.1016/0160-9327(92)90044-P

Ouyang S, Buell CR (2004) The TIGR Plant Repeat Databases: a collective resource for the identification of repetitive sequences in plants. Nucleic Acids Res 32:D360–D363. https://doi.org/10.1093/nar/gkh099

Pootakham W, Jomchai N, Ruang-areerate P, Shearman JR, Sonthirod C, Sangsrakru D, Tragoonrung S, Tangphatsornruang S (2015) Genome-wide SNP discovery and identification of QTL associated with agronomic traits in oil palm using genotyping-by-sequencing (GBS). Genomics 105(5–6):288–295. https://doi.org/10.1016/j.ygeno.2015.02.002

Raberchault H, Ahee J, Guenin G (1970) Colonies cellulaires et formes èmbryos in vitro a patir de cultures d'embryons de palmier a huile (*Elaeis guineensis* Jacq.) al'aide de cultures de tissue foliares. Comptes Rendus l'Académie des Sci Paris Série D283:1735–1737

Rao V, Kushairi A (1999) Quality of oil palm planting material. In: Rajanaidu N, Jalani BS (eds) 1996 seminar on sourcing of oil palm planting materials for local and overseas joint ventures. PORIM, Bangi, pp 188–197

Rhoads A, Au KF (2015) PacBio sequencing and its applications. Genomics Proteomics Bioinformatics 13 (5):278-289. https://doi.org/10.1016/j.gpb.2015.08.002

Riju A, Chandrasekar A, Arunachalam V (2007) Mining for single nucleotide polymorphisms and insertions/deletions in expressed sequence tag libraries of oil palm. Bioinformation 2(4):128–131. https://doi.org/10.6026/97320630002128

Rival A (2000) Somatic embryogenesis in oil palm. In: Jain SM, Gupta PK, Newton RJ (eds) Somatic embryogenesis in woody plants, vol 6. Kluwer Academic Publishers, Massachusetts, pp 249–290

Rival A, Tregear J, Jaligot E, Morcillo F, Aberlenc F, Billotte N, Richaud F, Beule T, Borgel A, Duval Y (2001) Oil Palm Biotechnology at CIRAD. In: The 2001 PIPOC on cutting edge technologies for sustained competitiveness. Malaysian Palm Oil Board, pp 51–82

Rohani O, Sharifah S, Rafii M, Ong M, Tarmizi A, Zamzuri I (2000) Tissue culture of oil palm. In: Basiron Y, Jalani BS, Chan KW (eds) Advances in oil palm research. Malaysian Palm Oil Board, Bangi, pp 238–283

Rosli R, Amiruddin N, Ab Halim MA, Chan P-L, Chan K-L, Azizi N, Morris PE, Leslie Low E-T, Ong-Abdullah M, Sambanthamurthi R, Singh R, Murphy DJ (2018) Comparative genomic and transcriptomic analysis of selected fatty acid biosynthesis genes and CNL disease resistance genes in oil palm. PLoS ONE 13(4):e0194792. https://doi.org/10.1371/journal.pone.0194792

Royal Botanic Gardens Kew (1909) Bulletin of miscellaneous information. Springer on behalf of Royal Botanic Gardens. Royal Botanic Gardens, Kew

Sambanthamurthi R, Abrizah O, Umi Salamah R (1999) Biochemical factors that control oil composition in the oil palm. J Oil Palm Res (Special Issue):23–33

Sarpan N, Kok S-Y, Chai S-K, Fitrianto A, Nuraziyan A, Zamzuri I, Ong-Abdullah M, Ooi S-E (2015) A model for predicting flower development in *Elaeis guineensis* Jacq. J Oil Palm Res 27(4):315–325

Schweikert G, Zien A, Zeller G, Behr J, Dieterich C, Ong CS, Philips P, De Bona F, Hartmann L, Bohlen A, Kruger N, Sonnenburg S, Ratsch G (2009) mGene: Accurate SVM-based gene finding with an application to nematode genomes. Genome Res 19(11):2133–2143. https://doi.org/10.1101/gr.090597.108

Seng T-Y, Ritter E, Mohamed Saad SH, Leao L-J, Harminder Singh RS, Qamaruz Zaman F, Tan S-G, Syed Alwee SSR, Rao V (2016) QTLs for oil yield components in an elite oil palm (*Elaeis guineensis*) cross. Euphytica 212(3):399–425. https://doi.org/10.1007/s10681-016-1771-6

Shearman JR, Jantasuriyarat C, Sangsrakru D, Yoocha T, Vannavichit A, Tragoonrung S, Tangphatsornruang S (2013) Transcriptome analysis of normal and mantled developing oil palm flower and fruit. Genomics 101 (5):306–312. https://doi.org/10.1016/j.ygeno.2013.02.012

Sim S-C, Durstewitz G, Plieske J, Wieseke R, Ganal MW, Van Deynze A, Hamilton JP, Buell CR, Causse M, Wijeratne S, Francis DM (2012) Development of a large SNP genotyping array and generation of high-density genetic maps in tomato. PLoS ONE 7(7): e40563. https://doi.org/10.1371/journal.pone.0040563

Simão FA, Waterhouse RM, Ioannidis P, Kriventseva EV, Zdobnov EM (2015) BUSCO: assessing genome assembly and annotation completeness with single-copy orthologs. Bioinformatics 31(19):3210–3212. https://doi.org/10.1093/bioinformatics/btv351

Singh R, Soon-guan T, Panandam J, Rahman RA (2008a) Identification of cDNA-RFLP markers and their use for molecular mapping in oil palm (*Elaeis guineensis*). Asia Pacific J Mol Biol Biotechnol 16(3):53–63

Singh R, Zaki N, Ting N-C, Rosli R, Tan S-G, Low E-T, Ithnin M, Cheah S-C (2008b) Exploiting an oil palm EST database for the development of gene-derived SSR markers and their exploitation for assessment of genetic diversity. Biologia 63(2):227–235. https://doi.org/10.2478/s11756-008-0041-z

Singh R, Tan SG, Panandam JM, Rahman R, Ooi LC, Low E-TL, Sharma M, Jansen J, Cheah S-C (2009) Mapping quantitative trait loci (QTLs) for fatty acid composition in an interspecific cross of oil palm. BMC

Plant Biol 9(1):114. https://doi.org/10.1186/1471-2229-9-114

Singh R, Low E-TL, Ooi LC-L, Ong-Abdullah M, Ting N-C, Nagappan J, Nookiah R, Amiruddin MD, Rosli R, Manaf MAA, Chan K-L, Halim MA, Azizi N, Lakey N, Smith SW, Budiman MA, Hogan M, Bacher B, Van Brunt A, Wang C, Ordway JM, Sambanthamurthi R, Martienssen RA (2013a) The oil palm SHELL gene controls oil yield and encodes a homologue of SEEDSTICK. Nature 500(7462):340–344. https://doi.org/10.1038/nature12356

Singh R, Ong-Abdullah M, Low E-TL, Manaf MAA, Rosli R, Nookiah R, Ooi LC-L, Ooi S, Chan K-L, Halim MA, Azizi N, Nagappan J, Bacher B, Lakey N, Smith SW, He D, Hogan M, Budiman MA, Lee EK, DeSalle R, Kudrna D, Goicoechea JL, Wing RA, Wilson RK, Fulton RS, Ordway JM, Martienssen RA, Sambanthamurthi R (2013b) Oil palm genome sequence reveals divergence of interfertile species in Old and New worlds. Nature 500(7462):335–339. https://doi.org/10.1038/nature12309

Singh R, Low E-TL, Ooi LC-L, Ong-Abdullah M, Nookiah R, Ting N-C, Marjuni M, Chan P-L, Ithnin M, Manaf MAA, Nagappan J, Chan K-L, Rosli R, Halim MA, Azizi N, Budiman MA, Lakey N, Bacher B, Van Brunt A, Wang C, Hogan M, He D, MacDonald JD, Smith SW, Ordway JM, Martienssen RA, Sambanthamurthi R (2014) The oil palm VIRESCENS gene controls fruit colour and encodes a R2R3-MYB. Nat Commun 5(1):4106. https://doi.org/10.1038/ncomms5106

Singh R, Ong-Abdullah M, Low E, Nookiah R, Abd Manaf M, Sambanthamurthi R (2015) *S*ureSawit™ SHELL—a diagnostic assay to predict fruit forms of the oil palm. Oil Palm Bull 70:13–16

Siqueira JF, Fouad AF, Rôças IN (2012) Pyrosequencing as a tool for better understanding of human microbiomes. J Oral Microbiol 4(1):10743. https://doi.org/10.3402/jom.v4i0.10743

Smith SA, Donoghue MJ (2008) Rates of molecular evolution are linked to life history in flowering plants. Science 322(5898):86–89. https://doi.org/10.1126/science.1163197

Snyder EE, Stormo GD (1995) Identification of protein coding regions in genomic DNA. J Mol Biol 248:1–18. https://doi.org/10.1006/jmbi.1995.0198

Soares MB, Bonaldo MF, Jelene P, Su L, Lawton L, Efstratiadis A (1994) Construction and characterization of a normalized cDNA library. Proc Natl Acad Sci 91(2):9228–9232. https://doi.org/10.1073/pnas.91.20.9228

Solovyev V, Kosarev P, Seledsov I, Vorobyev D (2006) Automatic annotation of eukaryotic genes, pseudogenes and promoters. Genome Biol 7(Suppl 1):S10.1–S10.12. https://doi.org/10.1186/gb-2006-7-s1-s10

Soltani MF, Hadizadeh M, Soltani Banavandi MJ, Yazdizadeh A (2014) Bioinformatics comparison of codon usage of genes encoding phosphate transporter in terms of salt tolerance, day length, temperature and pollination in different plants. Int J Adv Biol Biomed Res 2(2):504–509

Stanke M, Waack S (2003) Gene prediction with a hidden Markov model and a new intron submodel. Bioinformatics 19:ii215–ii225. https://doi.org/10.1093/bioinformatics/btg1080

Stanke M, Diekhans M, Baertsch R, Haussler D (2008) Using native and syntenically mapped cDNA alignments to improve *de novo* gene finding. Bioinformatics 24(5):637–644. https://doi.org/10.1093/bioinformatics/btn013

Stein LD (2013) Using GBrowse 2.0 to visualize and share next-generation sequence data. Brief Bioinform 14(2):162–171. https://doi.org/10.1093/bib/bbt001

Sundararajan A, Dukowic-Schulze S, Kwicklis M, Engstrom K, Garcia N, Oviedo OJ, Ramaraj T, Gonzales MD, He Y, Wang M, Sun Q, Pillardy J, Kianian SF, Pawlowski WP, Chen C, Mudge J (2016) Gene evolutionary trajectories and GC patterns driven by recombination in *Zea mays*. Front Plant Sci 7:1433. https://doi.org/10.3389/fpls.2016.01433

Syed Alwee S, Seng T, Mohamed Saad S (2014) Applications of molecular markers for oil palm crop improvement. Planter 90(1057):281–291

Tatarinova TV, Alexandrov NN, Bouck JB, Feldmann KA (2010) GC3 biology in corn, rice, sorghum and other grasses. BMC Genom 11(1):308. https://doi.org/10.1186/1471-2164-11-308

Tatarinova T, Elhaik E, Pellegrini M (2013) Cross-species analysis of genic GC3 content and DNA methylation patterns. Genome Biol Evol 5(8):1443–1456. https://doi.org/10.1093/gbe/evt103

Tee S-S, Tan Y-C, Abdullah F, Ong-Abdullah M, Ho C-L (2013) Transcriptome of oil palm (*Elaeis guineensis* Jacq.) roots treated with *Ganoderma boninense*. Tree Genet Genomes 9(2):377–386. https://doi.org/10.1007/s11295-012-0559-7

Teh C-K, Ong A-L, Kwong Q-B, Apparow S, Chew F-T, Mayes S, Mohamed M, Appleton D, Kulaveerasingam H (2016) Genome-wide association study identifies three key loci for high mesocarp oil content in perennial crop oil palm. Sci Rep 6(1):19075. https://doi.org/10.1038/srep19075

Teh C-K, Muaz SD, Tangaya P, Fong P-Y, Mayes A-LO, Chew FT, Kulaveerasingam H, Appleton D (2017) Characterizing haploinsufficiency of *SHELL* gene to improve fruit form prediction in introgressive hybrids of oil palm. Sci Rep 7:3118. https://doi.org/10.1038/s41598-017-03225-7

Teoh WC, Cheah SC, Harikrishna K, Tan SH (2003) Isolation and characterization of Myb-related genes from oil palm (*Elaeis guineensis* Jacq.). J Plant Biol 46(2):95–104. https://doi.org/10.1007/BF03030437

Ter-Hovhannisyan V, Lomsadze A, Chernoff YO, Borodovsky M (2008) Gene prediction in novel fungal genomes using an *ab initio* algorithm with unsupervised training. Genome Res 18(12):1979–1990. https://doi.org/10.1101/gr.081612.108

Ting NC, Cheah SC, Ishak Z, Tan SG, Faridah QZ, Maizura I, Singh R (2006) Statistical mapping of quantitative trait loci controlling the time to first callusing in oil palm (*Elaeis guineensis* Jacq.) Tissue Culture. Pertanika J Trop Agric Sci 29(2006):35–45

Ting N-C, Zaki NM, Rosli R, Low E-TL, Ithnin M, Cheah S-C, Tan S-G, Singh R (2010) SSR mining in oil palm EST database: application in oil palm germplasm diversity studies. J Genet 89(2):135–145. https://doi.org/10.1007/s12041-010-0053-7

Ting N-C, Jansen J, Nagappan J, Ishak Z, Chin C-W, Tan S-G, Cheah S-C, Singh R (2013) Identification of QTLs associated with callogenesis and embryogenesis in oil palm using genetic linkage maps improved with SSR markers. PLoS ONE 8(1):e53076. https://doi.org/10.1371/journal.pone.0053076

Ting N-CC, Jansen J, Mayes S, Massawe F, Sambanthamurthi R, Ooi LCL, Chin CW, Arulandoo X, Seng T-YY, Alwee SSRS, Ithnin M, Singh R, Cheng-Li OL, Chin CW, Arulandoo X, Seng T-YY, Alwee SSRS, Ithinin M, Singh R (2014) High density SNP and SSR-based genetic maps of two independent oil palm hybrids. BMC Genom 15(1):309. https://doi.org/10.1186/1471-2164-15-309

Ting N-C, Yaakub Z, Kamaruddin K, Mayes S, Massawe F, Sambanthamurthi R, Jansen J, Low LET, Ithnin M, Kushairi A, Arulandoo X, Rosli R, Chan K-L, Amiruddin N, Sritharan K, Lim CC, Nookiah R, Amiruddin MD, Singh R (2016) Fine-mapping and cross-validation of QTLs linked to fatty acid composition in multiple independent interspecific crosses of oil palm. BMC Genom 17(1):289. https://doi.org/10.1186/s12864-016-2607-4

Tisné S, Denis M, Cros D, Pomiès V, Riou V, Syahputra I, Omoré A, Durand-Gasselin T, Bouvet J-M, Cochard B (2015) Mixed model approach for IBD-based QTL mapping in a complex oil palm pedigree. BMC Genom 16(1):798. https://doi.org/10.1186/s12864-015-1985-3

Tranbarger TJ, Dussert S, Joët T, Argout X, Summo M, Champion A, Cros D, Omore A, Nouy B, Morcillo F, Joet T, Argout X, Summo M, Champion A, Cros D, Omore A, Nouy B, Morcillo F (2011) Regulatory mechanisms underlying oil palm fruit mesocarp maturation, ripening, and functional specialization in lipid and carotenoid metabolism. Plant Physiol 156 (2):564–584. https://doi.org/10.1104/pp.111.175141

Tranbarger TJ, Fooyontphanich K, Roongsattham P, Pizot M, Collin M, Jantasuriyarat C, Suraninpong P, Tragoonrung S, Dussert S, Verdeil J-L, Morcillo F (2017) Transcriptome analysis of cell wall and NAC domain transcription factor genes during *Elaeis guineensis* fruit ripening: evidence for widespread conservation within monocot and eudicot lineages. Front Plant Sci 8:603. https://doi.org/10.3389/fpls.2017.00603

Ukoskit K, Chanroj V, Bhusudsawang G, Pipatchartlearnwong K, Tangphatsornruang S, Tragoonrung S (2014) Oil palm (*Elaeis guineensis* Jacq.) linkage map, and quantitative trait locus analysis for sex ratio and related traits. Mol Breed 33(2):415–424. https://doi.org/10.1007/s11032-013-9959-0

Van de Peer Y, Fawcett JA, Proost S, Sterck L, Vandepoele K (2009) The flowering world: a tale of duplications. Trends Plant Sci 14(12):680–688. https://doi.org/10.1016/j.tplants.2009.09.001

Wang H-C, Hickey DA (2007) Rapid divergence of codon usage patterns within the rice genome. BMC Evol Biol 7:S6. https://doi.org/10.1186/1471-2148-7-S1-S6

Wang X, Wu J, Liang J, Cheng F, Wang X (2015) Brassica database (BRAD) version 2.0: integrating and mining Brassicaceae species genomic resources. Database 2015:bav093. https://doi.org/10.1093/database/bav093

Weissman SM (1987) Molecular genetic techniques for mapping the human genome. Mol Biol Med 4:133–143

Wong G, Tan C, Soh A (1997) Large scale propagation of oil palm clones: experiences to date. Acta Hortic 447:649–658

Wong YC, Teh HF, Mebus K, Ooi TEK, Bin Kwong Q, Koo KL, Ong CK, Mayes S, Chew FT, Appleton DR, Kulaveerasingam H (2017) Differential gene expression at different stages of mesocarp development in high- and low-yielding oil palm. BMC Genom 18(1):470. https://doi.org/10.1186/s12864-017-3855-7

Wooi K (1990) Oil palm (*Elaeis guineensis* Jacq.): tissue culture and micropropagation. In: Bajaj YPS (ed) Biotechnology in agriculture and forestry 10: legumes and oilseed crops. Springer, Berlin, pp 569–592

Wooi KC (1995) Oil palm tissue culture-current practice and constraints. In: Rao V, Henson IE, Rajanaidu N (eds) 1993 ISOPB international symposium on recent developments in oil palm tissue culture and biotechnology. PORIM, Kuala Lumpur, pp 21–32

Xie M, Yu B (2015) siRNA-directed DNA Methylation in Plants. Curr Genomics 16(1):23–31. https://doi.org/10.2174/1389202915666141128002211

Xu L, Yuan Y, Zhang L, Wan L, Zheng Y, Zhou P, Li D (2011) Identification and characterization of differential gene expression in the mesocarp and kernel of oil palm nuts using suppression subtractive hybridization. Tree Genet Genomes 7(5):999–1010. https://doi.org/10.1007/s11295-011-0390-6

Yamagishi M, Yoshida Y, Nakayama M (2012) The transcription factor LhMYB12 determines anthocyanin pigmentation in the tepals of Asiatic hybrid lilies (*Lilium* spp.) and regulates pigment quantity. Mol Breed 30(2):913–925. https://doi.org/10.1007/s11032-011-9675-6

Yang X, Wang J (2016) Genome-wide analysis of NBS-LRR genes in sorghum genome revealed several events contributing to NBS-LRR gene evolution in grass species. Evol Bioinforma 12:9–21. https://doi.org/10.4137/EBO.S36433

Ying XU, Mural RJ, Ralph Einstein J, Shah MB, Uberbacher EC (1996) GRAIL: a multi-agent neural network system for gene identification. Proc IEEE 84 (10):1544–1551. https://doi.org/10.1109/5.537117

Zaki NM, Singh R, Rosli R, Ismail I (2012) *Elaeis oleifera* genomic-SSR markers: exploitation in oil palm germplasm diversity and cross-amplification in Arecaceae. Int J Mol Sci 13(4):4069–4088. https://doi.org/10.3390/ijms13044069

Zeven AC (1967) The semi-wild oil palm and its industry in Africa. Agric Res Rep 689:28–33

Zulkifli Y, Rajinder S, Mohd Din A, Ting NC, Rajanaidu N, Kushairi A, Musa B, Mohamad O, Ismanizan I (2014) Inheritance of SSR and SNP loci in an oil palm interspecific hybrid backcross (BC2) population. J Oil Palm Res 26:203–213

Molecular Evolution and Genetic Diversity of Oil Palm Based on Sequencing and Analysis with Molecular Markers

8

Parisa Azizi, Mohamed Musa Hanafi, Mohamad Osman, Mohd Rafii Yusop, Maizura Ithnin, Sima Taheri, and Mahbod Sahebi

Abstract

Molecular marker advancement and utilization to exploit and identify genetic diversity are the most key developments in molecular genetic studies. Molecular markers are the most common tool for exploring plant genetic diversity and area pre- requisite for gene mapping, segregation and genetic analysis, forensic assessment, phylogenetic analysis, and the achievement of many other biological goals. Despite the development of several molecular marker types that are regularly applied in plant breeding, majority of these markers are confined in their functions due to their limited accessibility and the cost of analyses carried out on a large scale. However, the advent of sequencing technologies, including next-generation sequencing (NGS) and genotyping by sequencing (GBS), has transformed plant breeding via SSR and SNP development. To date, different types of sequencing technologies have been generated and reviewed. This chapter provides insight into the spectrum of molecular marker-based detection methods and their role in determining the genetic diversity of oil palms based on sequencing.

P. Azizi · M. M. Hanafi (✉)
Laboratory of Plantation Science and Technology, Institute of Plantation Studies, Universiti Putra Malaysia, 43400 Serdang, Selangor, Malaysia
e-mail: mmhanafi@upm.edu.my

P. Azizi · M. M. Hanafi · M. R. Yusop · M. Sahebi
Laboratory of Climate-Smart Food Crop Production, Institute of Tropical Agriculture and Food Security, Universiti Putra Malaysia, Serdang, Selangor, Malaysia

M. M. Hanafi
Department of Land Management, Faculty of Agriculture, Universiti Putra Malaysia, 43400 Serdang, Selangor, Malaysia

M. Osman
Prime Minister's Department, Malaysian Industry-Government Group for High Technology (MIGHT), MIGHT Partnership Hub, Jalan Impact, 63000 Cyberjaya, Selangor, Malaysia

M. Ithnin
Advanced Biotechnology and Breeding Centre, Malaysian Palm Oil Board, No. 6, Persiaran Institusi, Bandar Baru Bangi, 43000 Kajang, Selangor, Malaysia

S. Taheri
Centre of Research in Biotechnology for Agriculture (CEBAR), University of Malaya, 50603 Kuala Lumpur, Malaysia

8.1 Introduction

Wide-ranging studies of the genetic differences in the oil palm germplasm would be valuable to determine whether taxonomic categorizations based on morphological characteristics expose patterns of genomic segregation. These investigations also help breeders to utilize genetic resources with shortened breeding processes for

© Springer Nature Switzerland AG 2020
M. Ithnin and A. Kushairi (eds.), *The Oil Palm Genome*, Compendium of Plant Genomes, https://doi.org/10.1007/978-3-030-22549-0_8

developing cultivars by providing useful information on the structures of populations, numbers of alleles, and diversity characteristics of germplasm (Constantin et al. 2017). Germplasm characterization based on molecular markers has presently gained importance due to the quick generation and excellence of data. Genetic diversity within and between plant populations has been assessed using different techniques, including (i) morphological and (ii) biochemical characterization (allozyme) in the pre-genomic era and (iii) molecular (DNA) marker analysis, particularly the use of simple sequence repeats (SSRs), inter-simple sequence repeats (ISSRs), and single nucleotide polymorphism (SNPs) in the post-genomic era (Govindaraj et al. 2015). Markers can display comparable types of inheritance, including dominant, recessive, and codominant, while other characteristics are monitored (Govindaraj et al. 2015) Homozygotes can be discerned from heterozygotes by using codominant markers, which are commonly more instructive than the dominant markers.

Over recent years, the cultivation of oil palm has been quickly expanded, and oil palm is currently the main source of the world supply of vegetable oils. The countries leading the production of palm oil include Indonesia, Malaysia, Thailand, and Nigeria, with Indonesia and Malaysia as the main exporters in the world (Ling 2019). The Malaysian Palm Oil Board (MPOB) collected extensive germplasms of two species, namely *Elaeisguineensis* (African palms) and *Elaeis oleifera* (American palms), which was maintained as ex situ collections in Johor, Malaysia (Hayati et al. 2004). Understanding the genetic structure of natural oil palm populations through evaluating the yield and genetic diversity is important for oil palm breeding programs. This information could also be helpful for sustained ex situ preservation of the germplasm. The *E. guineensis* germplasm has been characterized using molecular markers, including isozymes (Hayati et al. 2004), restriction fragment length polymorphisms (RFLPs) (Maizura et al. 2006), amplified fragment length polymorphisms (AFLPs) (Kularatne 2000), random amplified polymorphic DNA (RAPD) (Rajanaidu et al. 2006), and SSRs (Singh et al. 2008; Ting et al. 2010). In addition, characterization of the *E. oleifera* germplasm using RAPD (Barcelos et al. 2002) and SSR markers (Billotte et al. 2001; Ting et al. 2010; Zaki et al. 2010; Zaki et al. 2012) has been previously reported. On the other hand, SSR markers have been developed by increasing the number of available sequenced fragments of *E. oleifera* (Zaki et al. 2012) and have been used to recognize the genetics of different species. This chapter provides insight into the spectrum of molecular marker (e.g., SSRs and SNPs) contributions to the genetic diversity of oil palms.

The commercial *E. guineensis* planting materials, known as *tenera* (*T*), are derived from hybridization between *dura* (*D*) and *pisifera* (P). The *dura* has thick shell while the *pisifera* is shell-less (Beirnaert and Vanderweyen 1941). The *tenera* exhibits thinner shell than the *dura* with reasonably good oil yields. Historically, the oil palm industry in Southeast Asia expanded from four palms planted at Bogor Botanical Gardenin Java, Indonesia (Hartley 1988). These palms, known as Deli *dura* (Hardon and Thomas 1968) evolved into Ulu Remis (URD), Johore Labis (JLD), Elmina (E), and Banting (BD) populations in Malaysia. Intensive selection was applied and individuals that recorded good yields and combining ability with *pisifera* (e.g., *Yangambi*, *AVROS* (*Algemene Vereniging van Rubberplanters ter Oostkust van Sumatra*), La Me, Ekona) palms are used in seed production.

8.2 Available Molecular Marker Tools for Oil Palm Genetic Studies

Molecular markers are the most utilized genetic markers for genetic and molecular diversity analyses. Genetic variations caused by deletion, insertion, duplication, and inversion in the chromosome can be detected by these markers (Paterson 1996). Molecular markers are commonly classified into three clusters according to their technique of detection: polymerase chain reaction (PCR)-based (RAPDs, AFLPs, SSRs), hybridization-based (RFLPs), and DNA sequence-based markers (SNPs). The AFLP method uses a combination of PCR-based and RFLP markers (Mohan et al. 1997). To generate AFLP markers, PCR-amplified fragments are digested by particular restriction enzymes that digest DNA at or close to a particular recognition site in nucleotides. The SSR, sequence characterized amplified region (SCAR), sequence-tagged site (STS) and expressed sequence tag-simple sequence repeat (EST-SSR) markers are different classes of molecular markers that rely on the accessibility of short repeat sequences of oligonucleotides, while single nucleotide polymorphism (SNP) markers rely on single nucleotide base polymorphisms in the genome sequences of organisms.

8.2.1 Random-Amplified Polymorphic DNAs

Random-Amplified Polymorphic DNAs (RAPD) markers have been introduced as the first PCR-based markers and were detected from amplification of genomic DNA using random primers. The amplification process known as polymerase chain reaction (PCR) produce multiple copies of DNA strands (Welsh and McClelland 1990; Jacobson and Hedrén 2007). PCR-based marker systems are quicker and need minimum amount of DNA material.

8.2.2 Simple Sequence Repeats

The wealth of simple repeat sequences of DNA (e.g., di-, tri-, tetra- and penta-nucleotide repeats) throughout the oil palm genome was previously reported by Cheah et al. (1995). Construction of genomic libraries containing high percentages of Simple Sequence Repeats (SSRs) was not initially successful, and only one percent of clones is presented SSR sequences (Cheah and Ooi 1999). Subsequently, an enriched oil palm genomic library with SSR markers was constructed with 72% of clones presenting SSR sequences (Billotte et al. 2001).

Several advantages of microsatellites include codominant inheritance, high polymorphism, reproducibility, and poly-allelic nature in plant species. These make SSRs highly useful for breeders to study relationship between phenotypes and genotypes. However, the conventional method of library construction, cloning, and enrichment in SSRs detection is laborious and costly (Zane et al. 2002; Squirrell et al. 2003; Weising et al. 2005). The express sequence tag-SSRs (EST-SSRs) or genic microsatellites that characterize genic parts of the genome are molecular markers developed for putative functions derived from database searches and other in silico approaches. As of 2016, asmanyas 40,979 oil palm EST-SSRs are accessible on the National Center for Biotechnology Information (NCBI) website (Babu et al. 2019).

The EST markers have been used to study gene expression in oil palm (Jouannic et al. 2005). In this study, 2411 applicable ESTs were developed from five different cDNA libraries constructed from female and male shoot zygotic embryos, apices and inflorescences. Singh et al. (2008), reported a total of 5521 ESTs. Of these, dinucleotide repeats were the most frequent, followed by trinucleotide repeats. Primers were developed for 94 (69.1%) distinctive ESTs, comprising of 14 consensus and 80 singletons. In another study, Ting et al. (2010) detected 629 EST-SSRs containing 722 SSR markers for oil palm out of 10,258 unigenes with a variety of motifs. Of the total repeats, 45.6% were dinucleotide repeats, with 66.9% of the total being AG/CT motifs, 21.9% AT/AT, 10.9% AC/GT and 0.3% CG/CG. Trinucleotide repeats formed the second most frequent group (34.5%), with 23.3% of their total being AAG/CTT motifs, followed by AGG/CCT (13.7%), CCG/CGG (11.2%), AAT/ATT (10.8%), AGC/GCT (10.0%), ACT/AGT (8.8%), ACG/CGT (7.6%), ACC/GGT (7.2%), AAC/GTT (3.6%) and AGT/ACT (3.6%).

In the sequencing era, the release of whole genome sequences of *E. guineensis* enhanced the development of larger number of microsatellite markers and verification of their distribution over different chromosomes in the both *Elaeis* species. For 7256 identified microsatellites, 135 SSR primers showed cross-genus transferability. Of these, 16.7–93.3% revealed polymorphism, with average 73.7% (Xiao et al. 2016). In addition, based on de novo assembly, out of a total of 130,840 potential SSRs, 61 SSRs showed polymorphism among 11 lines derived from three different clones of oil palm (Taeprayoon et al. 2016). In a study using next-generation sequencing, ten mitochondrial SSR markers were derived from the mitochondrial genome of oil palm. Of these, nine were polymorphic and transferrable across 15 Arecaceae species (Uthaipaisanwong et al. 2017).

The development of ESR-SSR markers through transcriptome sequencing has been performed in many species including oil palms. A total of 182 EST-SSR markers were developed from transcriptome cDNA library constructed using cold stress tolerance *E. guineensis*palms (Xiao et al. 2014). With the whole genome database of oil palm and accessibility to large ESTs, a database for oil palm microsatellite markers has been developed. Using an existing EST oil palm sequence database, Babu et al. (2019), developed 3950 microsatellites covering one microsatellite marker in every 6.7 Kb of EST sequences. They found that tri- and tetranucleotide repeats were the most abundant followed by dinucleotides. Among 16 oil palm chromosomes, 245,654 SSR repeats were detected from whole genome-wide analysis, harboring 38,717 compound microsatellite repeats.

8.2.3 Single Nucleotide Polymorphisms

Identification of single nucleotide base differences and their order along DNA strands is possible through sequencing technologies such as NGS and genotyping by sequencing (GBS) (França et al. 2002). Single Nucleotide Polymorphisms (SNPs) are single-base substitutions that can occur at high level of frequencies near a targeted gene of interest. Abundant SNPs that uniformly distributed across a genome have recently been applied for establishing high-density genome-wide scans in humans and plants (Wang et al. 1998; Huang et al. 2010). Besides diversity studies, cultivar identification, phylogenetic analysis, and marker trait analysis, SNPs have also been applied for haplotyping systems for target genes or regions and map-based positional cloning (Varshney et al. 2008; Hiremath et al. 2012; Dhanapal et al. 2015).

Riju et al. (2007) extracted oil palm EST sequences from NCBI EST database (dbEST) and discovered 1180 SNP sites and 137 indel polymorphisms with a frequency of 1.36 SNPs/100 bp. Among six oil palm tissues used to generate the EST libraries, mesocarp, and zygotic embryo recorded the highest (2.91) and lowest (0.15) SNPs and indels per 100 bp, respectively.

To date, oil palm genotyping has been moderately conducted by SNP array. Oil palm breeding populations have been genotyped using 4.5 K custom Illumina SNP array (Ting et al. 2014). This array has been verified to be valuable and helpful in comparison and construction of linkage maps. However, an advanced density genotyping array is necessary to scan large genomes using different analyses, such as GWAS and linkage disequilibrium (LD). In a research carried out by Teh et al. (2016), 132 oil palm plants from 59 breeding origins were utilized in order to whole genome re-sequencing, resulted in construction of ~900 million raw reads and identification of 7.8 million SNPs. From these, a 200 K SNP array was designed to genotype 2045 ten era palms for GWAS analysis. These SNPs have been selected according to their shortrange LD distance, inherent with long breeding cycles and heterogeneous populations.

The possibility of SNP development from oil palm (*E. guineensis*) mesocarp tissue using two sets of transcriptome data obtained by 454 sequencing was reported by Pootakham et al. (2013). In this study, out of 823 identified putative SNP positions in silico, 46 biallelic SNPs were validated by Sanger sequencing technology (Pootakham et al. 2013). Later, Pootakham et al. (2015), discovered another 21,471 SNPs in oil palm using GBS method.

The Illumina HiSeq 2000 platform was used to develop 812 SNPs to study the genetic diversity of American oil palm (*E. oleifera*) populations (Pereira et al. 2015). In another study, 188 SNPs were discovered through Illumina NextSeq 500 technology for construction of a linkage map for QTL mapping of a hybrid oil palm (*tenera*) derived from Deli *dura* and *AVROS pisifera* parents (Bai et al. 2018). Jin et al. (2016) sequenced the genomes of 17 palms and an elite *dura* palm. Approximately, 10,971 scaffolds from the *dura* palm were successfully assembled with a length of 1.701 Gb, covering 94.49% of the oil palm genome. The number of predicted genes was 36,105. Additionally, 18.1 million SNPs were detected when the 17 additional palm trees were sequenced.

Numerous genetic diversity studies in oil palm were carried out using molecular markers, including isozymes, RAPDs, RFLPs, AFLPs, SSRs, and SNPs (Table 8.1). The outcomes from these works are further elaborated in the following sections.

Table 8.1 Publications describing genetic diversity studies carried out in oil palm populations

Marker type applied	Population	References
Isozymes	Pollen of seven accessions of *E. oleifera* and hybrid between *E. oleifera* and *E. guineensis*	Ataga and Fatokun (1989)
SSRs	194 oil palms from 49 populations	Bakoumé et al. (2015)
RFLP	11 oil palm germplasm collections	Maizura et al. (2001)
AFLP	687 accessions belonging to 11 African countries and Deli *dura*	Kularatne et al. (2001)
AFLP and RFLP	Within oil palm germplasm (both *E. oleifera* and *E. guineensis*)	Barcelos et al. (2002)
RAPD	*E. oleifera* accession collected from the Amazon forest	Moretzsohn et al. (2002)
RAPD	Five *dura* germplasm accessions	Mandal et al. (2004)
SSR	Elite oil palms from Nigeria and selected breeding and germplasm materials from Malaysia	Okoye et al. (2016)
SSR	*E. oleifera* and *E. guineensis*	Zaki et al. (2012)
SSR	NIFOR oil palm main breeding parent genotypes	Okoye et al. (2016)
SSR	121 breeding plants from three different populations in Thailand	Taeprayoon et al. (2015)
RAM	51 oil palm genotypes from the Congo (*E. guineensis* Jacq.)	Cardona et al. (2018)
EST, SSR	Genotypes from the world oil palm genetic resources (*E. guineensis* Jacq.)	Babu et al. (2019)
SSR	Dura, Algemene Vereniging van Rubber-planters ten Oostkust van Sumatra (AVROS) and Genting AgTech Sdn Bhd (GAT) germplasm materials (*E. guineensis* Jacq.) from Malaysia	Chun et al. (2018)
SSR	NIFOR *dura* × *tenera* oil palm progenies	Okoye et al. (2018)
SSR	Eight oil palm genotypes (*E. guineensis* Jacq.)	Bhagya et al. (2018)
SSR	*Tenera* hybrid populations	Khomphet et al. (2018)
SSR	*Dura* × *pisifera* population	Solin and Toruan-Mathius 2014)
SNP	*Elaeis guineensis*, 84 breeding lines, 88 introgressed palms, and interspecific cross	Nugroho et al. (2019)
SNP	*Elaeis guineensis*from different geographical regions (including, China, Malaysia, Costa Rica, and Africa)	Xia et al. (2019)

RAM random amplified microsatellite

8.3 Molecular Marker Applications in Oil Palm Genetic Diversity Studies

8.3.1 Diversity Analysis Using RAPDs and AFLPs

Application of RAPD markers for investigation of oil palm genetic diversity was reported by Shah et al. (1994). Using 20 primers, high variation was detected across oil palm accessions originated from Nigeria, Cameroon, Zaire, and Tanzania. RAPD markers (387 arbitrary primers) have also been applied to study oil palm somaclonal variation (Rival et al. 1998). However, results demonstrated that RAPDs are not applicable neither in understanding oil palm somaclonal variation that affects oil palm productivity nor in differentiating the original oil palm and their regenerated clones.

Assessment of oil palm diversity using AFLP markers showed that crosses derived from African germplasms were more diverse than those produced from African and Deli palms (Purba et al. 2000). AFLPs confirmed the genetic structure revealed by RFLP where *E. oleifera* accessions were grouped according to geographical origins, namely Brazil, French Guyana/Surinam, Peru, and north of Colombia/Central America. Both markers showed that genetic divergence between the two *Elaeis* species is of the same magnitude as that among provenances of *E. oleifera* (Barcelos et al. 2002). Subsequently, 30 RAPD markers were used to determine DNA polymorphisms among three oil palm (*E. guineensis*) varieties, *pisifera, dura* and *tenera*. Of these, 26 primers were able to distinguish the varieties (Sathish and Mohankumar 2007). Jayanthi et al. (2008), applied RAPD markers to determine the genetic uniformity of oil palm clones. Out of ten RAPD primers used, six primers gave monomorphic bands. The similarity percentage ranged between 85 and 100% indicating that the clones are quite similar to each other. However, the dendrogram revealed that there was no absolute similarity among the clones as clones from one source were clustered with clones from other sources.

8.3.2 Simple Sequence Repeats

Genetic diversity characterization of oil palm populations by means of SSR markers were reported since 2001 (Billotte et al. 2001; Billotte et al. 2005; Bakoumé et al. 2007; Singh et al. 2008; Putri et al. 2010; Ting et al. 2010; Abdullah et al. 2011; Ajambang et al. 2012; Arias et al. 2012; Solin and Toruan-Mathius 2014; Bakoumé et al. 2015). These indicated the suitability of SSRs for genetic studies in *E. guineensis* as well as its related species. The transferability of SSRs across taxa allows researchers to apply currently available SSRs across various species and genus. This greatly saves time and cost especially for underutilized species where financial support and resources to pursue genetic studies are limited.

Bakoumé et al. (2007), studied the allelic variations among oil palm populations from different African origins using microsatellite markers. Results revealed several distinctive and rare alleles in natural oil palm populations from countries located in West Africa that experience lower rainfall and longer dry seasons. Further conservation of selected 23 populations that possessed unique alleles and high polymorphisms was recommended to ensure access for future breeding purposes. Ajambang et al. (2012), investigated the level of genetic variation among 39 accessions of wild oil palm population collected from seven regions in Cameroon using 16 SSR loci. The results indicated 96.43% polymorphic loci and a polymorphic information content of 0.597, indicating high level of genetic variation. The genetic variability within was higher than between populations thus more individuals per populations should be selected for future conservation and breeding programs.

Germplasm materials of *E. guineensis* collected from eleven African countries that are maintained by MPOB were evaluated using ten EST-SSR markers (Zulkifli et al. 2009). The EST-SSR alleles revealed 100% polymorphism indicating high genetic diversity among the populations. Out of 46 alleles detected across the germplasms, three were found to be rare alleles. Populations from Nigeria exhibited maximum number of alleles per locus, highest number of rare alleles, high percentage of polymorphic loci and heterozygosity which signified Nigeria germplasm as a center of diversity of wild oil palm. The germplasm collection was divided into three different clusters based on a constructed dendrogram. Cluster 1 consisted of Cameroon, Nigeria, Angola, Tanzania, Sierra Leone, Zaire, Ghana, and Guinea germplasms; cluster 2 comprised Senegal and Gambia germplasms; and the Madagascar germplasm was grouped in cluster 3. This study demonstrated Madagascar as a distinctive population compared to other germplasms originating from African mainland.

Babu et al. (2019), downloaded an oil palm EST database and developed 3950 SSR primer pairs. In addition, using oil palm genome sequence, the research group identified 38,717 genome-wide microsatellite repeats across 16 chromosomes. Some ten EST-SSRs and five genomic SSRs were applied for genetic diversity study across 100 oil palm germplasm accessions. The estimates of PIC, gene diversity, expected, and observed heterozygosity showed that genomic SSRs were more polymorphic than the EST-SSRs. The phylogenetic trees revealed dissimilar groupings probably due to the differences in the genomic region covered by each SSR type.

The genetic diversity of 76 accessions of oil palm collected from seven African countries and the standard *Deli dura* population was evaluated using 10 developed EST-SSR primers. The average number of observed and effective alleles was 2.56 and 1.84, respectively. The average polymorphic information content (PIC) was 0.53, while a considerable level of genetic diversity was confirmed by the genetic differentiation (FST) of 0.2492. Additionally, a significant relationship was discovered between genetic distance and geographic locations of the studied populations using an unweighted pair-group method with arithmetic mean (UPGMA) analysis. Higher diversity was observed among germplasm resources in populations from Congo, Cameroon, and Nigeria than in those of *Deli dura*, indicating that they are potent in oil palm breeding programs (Singh et al. 2008).

Elaeis oleifera natural accessions assembled from South American countries were characterized using SSRs and agro-morphological traits (Arias et al. 2015). Phenotypic trait studies illustrated significant variations within countries and geographical areas especially in terms of yield and bunch components. SSR screening revealed allele-specific-country. Cluster analysis of molecular data produced four genetic groups, which matched with the geographical distribution of the accessions. The accessions exhibited genetic structure and phenotypic diversity which are useful for development of strategy for breeding programs and core collection.

Genetic variation in the *tenera* hybrid progenies provides an effective direction in oil palm breeding programs, especially for the development of high-yielding *tenera* varieties. Khomphet et al. (2018), investigated the genetic diversity of 12 *tenera* hybrid populations based on morphology and SSR markers. Among them, P1, P4, P8, and P13 showed morphological and genetic variation indicating high potential for adaptation and yield stability. From principal component analysis (PCA), progenies with large PC1 and PC2 components were the most potent and highly productive. Norziha et al. (2008), estimated the genetic distance between parent palms (15 *dura* and 4 *pisifera*) and their progenies (16 $D \times P$) using nine microsatellite markers. A total of 29 polymorphic bands were generated with genetic distances between progenies ranging from 0.089 to 0.313.

Genetic similarity among 9 $D \times P$ crosses from different commercial seed production companies based in Costa Rica, Colombia, Malaysia, and France was estimated using 17 SSR primers (Arias et al. 2012). Genetic similarity coefficients ranged between 66 and 76% indicating considerable levels of differentiation among the crosses.

Results also showed that the crosses underwent significant reduction in the number of alleles when compared to wild oil palm populations. The dendrogram constructed showed two genetic groups. The first group contained $D \times P$ crosses produced in France and Colombia whereas the second group consisted of crosses generated in Malaysia, France, Costa Rica, and Colombia. Although the parental palms used to create the crosses originated from common sources, Arias et al. (2012), postulated that different sets of alleles were fixed due to different selection and breeding methods imposed by each company. This explained the formation of two genetic groups in the dendrogram.

Nine SSRs have been used by Okoye et al. (2016), to evaluate the genetic relationship among elite $D \times P$ oil palm populations produced in Nigeria and Malaysia. The populations exhibited high percentage of polymorphic loci (83.3%) with average polymorphic information content (PIC) of 0.7325. Compared to Malaysia populations, the Nigerian populations showed high genetic diversity as well as unique features (e.g., high bunch production and low stem increment). These highlighted the potential used of the Nigerians for improvement of plant architecture and yields.

Chun et al. (2018), estimated the genetic diversity in $D \times P$ populations and determined the levels of inbreeding using CIRAD's SSR markers. Genotyping using 32 SSR markers produced 230 alleles within the 17 progenies of oil palm. The number of alleles scored ranged between 4 and 11. The average number of alleles per locus was 7.1875. The observed heterozygosity (Ho) and expected heterozygosity (He) were 0.5270 and 0.7063, respectively. A dendrogram revealed that the 17 populations were clustered into three main clusters according to the genetic sources, namely Deli *dura, AVROS,* and *Ekona* populations. The diversity revealed via SSR showed that genetic variability still persists in some of the progenies.

The genetic diversity of ten *dura* × *tenera* oil palm crosses developed in the Nigerian Institute for Oil Palm Research (NIFOR) were evaluated using five agronomic markers and 16 SSRs (2018). Phenotypically, high level of variation was observed in terms of oil-to-mesocarp ratio (5.6–40.5%) and bunch number. These crosses also showed 100% polymorphism using SSR markers with medium genetic diversity (He = 0.661, Ho = 0.580) despite shared ancestry of the parents.

In another study, 110 SSR markers were used to screen genotypes of oil palm height phenotypes (Bhagya et al. 2018). Of these, 42 SSRs showed polymorphism, with two to six alleles per primer were observed. One of the primers, namely mEgCIR0779, had the highest polymorphic information content (PIC) (0.76) and genetic diversity (0.79) while mEgCIR3288 had the lowest (0.11 and 0.12, respectively). The polymorphic markers detected in the study are useful in genetic mapping studies of oil palm.

Putri et al. (2010), screened seven *pisifera* populations from AVROS, Nigeria, Ekona, Ghana, Dami, La Me and Yangambiorigins using SSRs. A total of 163 alleles were observed in the *pisifera* collections, with an average of 8.2 alleles per locus. Principal Coordinates Analysis (PCoA) showed four groups. The first and second groups, respectively, consisted of Yangambi and AVROS accessions. The third group was formed by Ekona accessions while the fourth group contained accessions from Dami, La Mé, Ghana, and Nigeria. Some outliers (two Nigerian and one Yangambi palm) were observed which may be due to mislabeling. Exclusion of such palms in breeding programs was recommended.

Tranbarger et al. (2011), screened elite breeding parents and their progenies using 289 EST-SSR markers. From these, 230 produced amplified products, with 88 SSRs revealed polymorphism. Further annotation of the EST-SSRs showed their relations with transcriptional and post-transcriptional systems. Ting et al. (2010), designed 405 primers, of which 105 primers were used for genotyping 105 *E. guineensis* and 30 *E. oleifera* accessions. A total of 101 alleles was observed and fourteen primers were polymorphic in at least one germplasm. High percentage of alleles (78.0%) was discovered either from *E. guineensis* or *E. oleifera* and these enhanced the power for distinguishing the two species. The transferability of the markers was further investigated across other palm taxa, namely *Cocos nucifera* and other six exotic palm species. PCR products amplified in *E. guineensis*, *E. oleifera*, *C. nucifera,* and *Jessiniabataua* were cloned and sequenced. The SSR sites and flanking regions showed mutations within SSRs after sequence alignment. Results also showed that *C. nucifera* is more similar to oil palm than *J. bataua*, which matched with the outcomes from genetic taxonomic grouping (Ting et al. 2010).

A number of authors have demonstrated the importance of the SSR technique for genetic fingerprinting in oil palm (Singh et al. 2007; Norziha et al. 2008; Thawaro and Te-chato 2010; Bakoumé 2011; Chee et al. 2015; Hama-Ali et al. 2015). Chee et al. (2015), screened 33 oil palm genotypes using 17 highly polymorphic SSR markers and suggested that, clones, crosses, and sibs within crosses could be categorized in separate groups using their DNA fingerprints. However, they subsequently reported that palms within clones could not be separated in an individual cluster. Moreover, four related populations, including Deli, Dumpy Deli × Yangambi × AVROS, Dumpy Deli × AVROS, Dumpy Deli × AVROS × La Me were tested using 11 SSR markers and clearly separated. In another finger printing study of oil palms populations, thirty polymorphic SSR markers were used to identify correct parents and progenies among half-sib families from controlled crosses of *dura* (FD8, FD6, FD10 and FD1/224 as female parents) and*pisifera* (FP1/10and FP1/28 as male parents). Sixteen loci were found to be adequate for correct assignment of parent-offspring and identification of illegitimate palms. Thawaro and Te-chato (2010), verified the legitimacy of *tenera* hybrids in six $D \times P$ crosses using eight SSR primers. The results revealed that primers mEgCIR008 offer an obvious DNA pattern that could be applied for hybrid verification in the crosses. Using 10 SSRs, Bakoumé (2011) successfully distinguished clones produced from different tissue donors. This indicated the utility of SSRs for routine verification of clone-to-donor identity. Similarly, Rajinder et al. (2007) reported a reasonable division of six clone-donor sets using five SSR loci. These researchers recommended that a minimum of five SSR markers are sufficient for clonal identification and detection of off-types.

8.3.3 Single Nucleotide Polymorphism

To estimate the genetic diversity of 219 oil palms from two natural Angolan populations, 62 SNP primers were developed from genomic sequences of oil palm (Ong et al. 2015). Using cleaved amplified polymorphic sequences (CAPS) method, nine SNPs were instructive among the two populations, with an average allele frequency of 0.693. The average observed and expected heterozygosities were 0.400 and 0.398, respectively. The PIC varied from 0.223 to 0.375, with an average of 0.315. All the loci adhered to Hardy–Weinberg equilibrium, and no rare allele was detected. Based on UPGMA analysis, 219 oil palms grouped into two categories. However, the grouping of the palms did not match with geographic distribution. The genetic variation of the populations was contributed mainly by differentiation within population (93%).

High-density SNPs are highly preferential markers for estimating genetic diversity, population structure and constructing high density genetic maps. Xia et al. (2019), applied single locus amplified fragment sequencing (SLAF-seq) technology to develop genome-wide SNPs in oil palm (*E. guineensis*). Of 1261,501 SNPs, only 17.81% were located within the genic region. Further, there was positive correlation between the distribution of SNPs and transposable elements (TEs) throughout the genome. The SNPs were used to study genetic diversity of oil palm populations in China, Malaysia, Africa, and Costa Rica. Population structure analysis resulted in five clusters. Oil palms from Malaysia grouped together, along with those from Costa Rica. However, the oil palms from Africa and China were mixed. The estimated linkage disequilibrium was 14.516 kb at $r^2 = 0.1$, indicating rapid decay within short physical distance which helps in approximating the number of markers needed for genome-wide genetic studies.

In another study, Nugroho et al. (2019), carried out paired-end Double Digest Restriction Associated DNA (ddRAD) sequencing on 236 palms from diverse genetic backgrounds to discover high quality SNPs and allelic variation. A sizeable amount of 195.62 Gb data was generated and 8189 SNPs were discovered within annotated genic regions. Approximately 29% of loci displayed moderate to high polymorphism based on in silico prediction of amino acid alterations. High fixation index and low observed heterozygosity were detected among samples from Angola and Deli populations indicating inbreeding. Weak population structure was obtained which explained the lack of ability to differentiate core breeding materials according to their genetic backgrounds.

SNPs detected from sequencing technology was applied to uncover genetic diversity of *teneras*, *duras*, *pisiferas,* and hybrids of *E. oleifera* and *E. guineensis* originating from Indonesia, Nigeria, Ghana, Angola, and Costa Rica (Jin et al. 2016). Low genetic diversity was detected, possibly due to extensive selection and breeding applied on the palms. Introduction of new genetic source was recommended to ensure selection gains in the future.

8.4 Conclusion

Among the molecular techniques currently available for assessment of genetic diversity, SSR and SNP markers more commonly used owing to their richness in polymorphism and stability. Both markers are amendable to high throughput platforms which enables genotyping of large number of markers across thousands of samples at reasonable cost. The amount of data generated fulfills the statistics and accuracy of any genetic analysis which improve the comprehension on oil palm. Both, SSR and SNP markers have been successfully applied in assessing genetic diversity of commercial and natural oil palm populations worldwide. Combined knowledge on morphological and molecular marker variation provides better insights and informed choice to oil palm breeders in prioritizing genetic materials for breeding and future conservation.

Acknowledgements The authors wish to acknowledge Universiti Putra Malaysia (UPM) for support.

References

Abdullah N, Yusop MR, Ithnin M, Saleh G, Latif M (2011) Genetic variability of oil palm parental genotypes and performance of its' progenies as revealed by molecular markers and quantitative traits. CR Biol 334:290–299

Ajambang W, Asmono D, Toruan N (2012) Microsatellite markers reveal Cameroon's wild oil palm population as a possible solution to broaden the genetic base in the Indonesia-Malaysia oil palm breeding programs. Afr J Biotech 11:13244–13249

Arias D, Montoya C, Rey L, Romero H (2012) Genetic similarity among commercial oil palm materials based on microsatellite markers. Agronomía Colombiana 30:188–195

Arias D, González M, Romero H (2015) Genetic diversity and establishment of a core collection of oil palm (*Elaeis guineensis* Jacq.) based on molecular data. Plant Genet Resour 13:256–265

Ataga C, Fatokun C (1989) Disc polyacrylamide gel electrophoresis of pollen proteins in the oil palm (Elaeis). Euphytica 40:83–88

Babu K et al (2019) Development and validation of whole genome-wide and genic microsatellite markers in oil palm (*Elaeis guineensis* Jacq.): first microsatellite database (OpSatdb). Sci Rep 9:1899

Bai B et al (2018) Developing genome-wide SNPs and constructing an ultrahigh-density linkage map in oil palm. Sci Rep 8:691

Bakoumé C et al (2011) DNA sequence-based markers for verification of ramet-to-ortet relationship in oil palm (*Elaeis guineensis* Jacq.). Am J Plant Sci 2:539

Bakoumé C, Wickneswari R, Rajanaidu N, Kushairi A, Amblard P, Billotte N (2007) Allelic diversity of natural oil palm (*Elaeis guineensis* Jacq.) populations detected by microsatellite markers: implications for conservation. Plant Genet Resour 5:104–107

Bakoumé C, Wickneswari R, Siju S, Rajanaidu N, Kushairi A, Billotte N (2015) Genetic diversity of the world's largest oil palm (*Elaeis guineensis* Jacq.) field genebank accessions using microsatellite markers. Genet Resour Crop Evol 62:349–360

Barcelos E, Amblard P, Berthaud J, Seguin M (2002) Genetic diversity and relationship in American and African oil palm as revealed by RFLP and AFLP molecular markers. Pesquisa Agropecuária Brasileira 37:1105–1114

Beirnaert ADF, Vanderweyen R (1941) Contribution à l'étude génétique et biométrique des variétés d'*Elaeis guineensis* Jacquin. East African Standard

Bhagya H, Babu BK, Naika MB, Mathur R, Gangadharappa P, Satisha D, Naik R (2018) Identification and utilization of polymorphic SSR markers for genetic diversity studies in oil palm. Int J Curr Microbiol App Sci 7:333–341

Billotte N, Risterucci A-M, Barcelos E, Noyer J-L, Amblard P, Baurens F-C (2001) Development, characterisation, and across-taxa utility of oil palm (*Elaeis guineensis* Jacq.) microsatellite markers. Genome 44:413–425

Billotte N et al (2005) Microsatellite-based high density linkage map in oil palm (*Elaeis guineensis* Jacq.). Theor Appl Genet 110:754–765

Cardona CCC, Coronado YM, Conronado ACM, Ochoa I (2018) Genetic diversity in oil palm (*Elaeis guineensis* Jacq.) using RAM (random amplified microsatellites). Bragantia 77:546–556

Cheah S, Ooi L (1999) Development of genetic markers for the oil palm based on simple sequence repeat (SSR) DNA. In: Colloquium on advances in oil palm research under IRPA-funded programmes in the 7th Malaysian Plan, pp 1–2

Cheah SC, Ooi LCL, Rahimah AR (1995) Polymorphic DNA in oil palm amplified by simple sequence repeat (SSR) primed polymerase chain reactions. Paper presented at the In The VII National Biotechnology Seminar (Seminar Bioteknologi Kebangsaan ke VII), Langkawi Island, Malaysia, 20–22 November

Chee WW, Jit TC, Kien WC, Mayes S, Singh R, Chin S (2015) Development of an effective SSR-based fingerprinting system for commercial planting materials and breeding applications in oil palm. J Oil Palm Res 27:113–127

Chun TC et al (2018) Genetic diversity and inbreeding level in deli dura and avros advanced breeding materials in oil palm (*Elaeis guineensis* Jacq.) using microsatellite markers. J Oil Palm Res 30:366–379

Constantin M, Ridwani S, Syukur M, Suwarno WB, Godswill N-N (2017) Genetic Diversity and Interrelationship among Some Dura × Tenera Oil Palm (*Elaeis guineensis* Jacq.) Genotypes in Cameroon. J Agric Sci Technol 7:81–90

Dhanapal AP et al (2015) Genome-wide association analysis of diverse soybean genotypes reveals novel markers for nitrogen traits. Plant Genome 8

França LT, Carrilho E, Kist TB (2002) A review of DNA sequencing techniques. Q Rev Biophys 35:169–200

Govindaraj M, Vetriventhan M, Srinivasan M (2015) Importance of genetic diversity assessment in crop plants and its recent advances: an overview of its analytical perspectives. Genet Res Int 2015:1–14

Hama-Ali EO, Alwee SSRS, Tan SG, Panandam JM, Ling HC, Namasivayam P, Peng HB (2015) Illegitimacy and sibship assignments in oil palm (*Elaeis guineensis* Jacq.) half-sib families using single locus DNA microsatellite markers. Mol Biol Rep 42:917–925

Hardon J, Thomas R (1968) Breeding and selection of the oil palm in Malaya. Oléagineux 23:85–90

Hartley C (1988) The oil palm, 3rd edn. Longman, London

Hayati A, Wickneswari R, Maizura I, Rajanaidu N (2004) Genetic diversity of oil palm (*Elaeis guineensis* Jacq.) germplasm collections from Africa: implications for improvement and conservation of genetic resources. Theor Appl Genet 108:1274–1284

Hiremath PJ et al (2012) Large-scale development of cost-effective SNP marker assays for diversity assessment and genetic mapping in chickpea and comparative mapping in legumes. Plant Biotechnol J 10:716–732

Huang X et al (2010) Genome-wide association studies of 14 agronomic traits in rice landraces. Nat Genet 42:961

Jacobson A, Hedrén M (2007) Phylogenetic relationships in Alisma (Alismataceae) based on RAPDs, and sequence data from ITS and trnL. Plant Syst Evol 265:27–44

Jayanthi M, Mandal P, Sujatha G, Sri K, Rao G, Sunitha B, Babu M (2008) Simple sequence repeats and RAPD primers for assessment of genetic uniformity among the field planted clones of oil palm. J Plant Crops 36:235–238

Jin J et al (2016) Draft genome sequence of an elite Dura palm and whole-genome patterns of DNA variation in oil palm. DNA Res 23:527–533

Jouannic S et al (2005) Analysis of expressed sequence tags from oil palm (*Elaeis guineensis*). FEBS Lett 579:2709–2714

Khomphet T, Eksomtramage T, Duangpan S (2018) Genetic variation of improved oil palm Tenera hybrid populations using morphological and SSR markers. Songklanakarin J Sci Technol 40

Kularatne R (2000) Assessment of genetic diversity in natural oil palm (*Elaeis guineensis* Jacq.) populations using amplified fragment length polymorphism markers. Ph. D. Dissertation, Universiti Kebangsaan Malaysia, Kuala Lumpur, Malaysia

Kularatne R, Shah F, Rajanaidu N (2001) The evaluation of genetic diversity of Deli dura and African oil palm germplasm collection by AFLP technique

Ling AH (2019) Global palm oil trade—prospect and outlook. Paper presented at Malaysia—China Business Forum, Kuala Lumpur, 4th March 2019

Maizura I, Cheah S, Rajanaidu N (2001) Genetic diversity of oil palm germplasm collections using RFLPs. In: Cutting-edge technologies for sustained competitiveness: Proceedings of the 2001 PIPOC International Palm Oil Congress, Agriculture Conference, Kuala Lumpur, Malaysia, 20–22 August 2001, Malaysian Palm Oil Board (MPOB), pp 526–535

Maizura I, Rajanaidu N, Zakri A, Cheah S (2006) Assessment of genetic diversity in oil palm (*Elaeis guineensis* Jacq.) using restriction fragment length polymorphism (RFLP). Genet Resour Crop Evol 53:187–195

Mandal PK, Malliah P, Sireesha K, Shamila S, Aruna C (2004) The use of RAPD markers for molecular characterization of oil palm (*Elaeis guineensis* Jacq.) germplasm. J Plant Crops 32:131–133

Mohan M, Nair S, Bhagwat A, Krishna T, Yano M, Bhatia C, Sasaki T (1997) Genome mapping, molecular markers and marker-assisted selection in crop plants. Mol Breeding 3:87–103

Moretzsohn MdC, Ferreira M, Amaral Z, Coelho PJdA, Grattapaglia D, Ferreira ME (2002) Genetic diversity of Brazilian oil palm (Elaeis oleifera HBK) germplasm collected in the Amazon Forest. Euphytica 124:35

Norziha A, Rafii M, Maizura I, Mohd Din A (2008) Genetic diversity among oil palm parental genotypes revealed by microsatellite polymorphism and its relationship to progeny performance

Nugroho Y et al (2019) Genome-wide SNP-discovery and analysis of genetic diversity in oil palm using double digest restriction site associated DNA sequencing. In: IOP conference series: earth and environmental science, vol 1. IOP Publishing, p 012041

Okoye M, Bakoumé C, Uguru M, Singh R, Okwuagwu C (2016) Genetic relationships between elite oil palms from Nigeria and selected breeding and germplasm materials from Malaysia via Simple Sequence Repeat (SSR) Markers. J Agric Sci 8

Okoye MN, Bakoume C, Uguru MI, Singh R (2018) Genetic diversity and relatedness of oil palm progenies determined by microsatellite and agronomic markers. Afr J Biotech 17:614–625

Ong P, Maizura I, Abdullah N, Rafii M, Ooi L, Low E, Singh R (2015) Development of SNP markers and their application for genetic diversity analysis in the oil palm (*Elaeis guineensis*). Genet Mol Res 14:12205–12216

Paterson AH (1996) Genome mapping in plants. RG Landes Co.

Pereira V, Leao A, Forimighieri E, Souza Junior M, Rios SdA, Alves A (2015) Molecular characterization and genetic structure of american oil palm (Elaeis oleifera) based on genome-wide SNP markers. In: Embrapa Agroenergia-Resumo em anais de congresso (ALICE). Encontro De Pesquisa E Inovação Da Embrapa Agroenergia. Brasília, DF. Anais… Brasília, DF: Embrapa Agroenergia

Pootakham W, Uthaipaisanwong P, Sangsrakru D, Yoocha T, Tragoonrung S, Tangphatsornruang S (2013) Development and characterization of single-nucleotide polymorphism markers from 454 transcriptome sequences in oil palm (*Elaeis guineensis*). Plant Breeding 132:711–717

Pootakham W et al (2015) Genome-wide SNP discovery and identification of QTL associated with agronomic traits in oil palm using genotyping-by-sequencing (GBS). Genomics 105:288–295

Purba AR, Noyer J-L, Baudouin L, Perrier X, Hamon S, Lagoda P (2000) A new aspect of genetic diversity of Indonesian oil palm (*Elaeis guineensis* Jacq.) revealed by isoenzyme and AFLP markers and its consequences for breeding. Theor Appl Genet 101:956–961

Putri LAP, Rivallan R, Puspitaningrum Y, Perrier X, Asmono D, Billotte N (2010) Allelic diversity of 22 Sampoerna Agro's oil palm pisifera based on microsatellite markers

Rajanaidu N, Maizura I, Cheah S (2006) Screening of oil palm natural populations using RAPD and RFLP molecular markers. Malaysian Palm Oil Board (MPOB)

Riju A, Chandrasekar A, Arunachalam V (2007) Mining for single nucleotide polymorphisms and insertions/deletions in expressed sequence tag libraries of oil palm. Bioinformation 2:128

Rival A, Bertrand L, Beulé T, Combes M-C, Trouslot P, Lashermes P (1998) Suitability of RAPD analysis for the detection of somaclonal variants in oil palm (*Elaeis guineensis* Jacq.). Plant Breeding 117:73–76

Sathish D, Mohankumar C (2007) RAPD markers for identifying oil palm (*Elaeis guineensis* Jacq.) parental varieties (dura & pisifera) and the hybrid tenera

Shah F, Rashid O, Simons A, Dunsdon A (1994) The utility of RAPD markers for the determination of genetic variation in oil palm (*Elaeis guineensis*). Theor Appl Genet 89:713–718

Singh R, Nagappan J, Tan SG, Panandam JM, Cheah SC (2007) Development of simple sequence repeat (SSR) markers for oil palm and their application in genetic mapping and fingerprinting of tissue culture clones. APJMBB 15:121–131

Singh R, Nagappan J, Tan S-G, Panandam JM, Cheah SC (2007) Development of simple sequence repeat (SSR) markers for oil palm and their application in genetic mapping and fingerprinting of tissue culture clones. Asia Pac J Mol Biol Biotechnol 15:121–131

Singh R et al (2008) Exploiting an oil palm EST database for the development of gene-derived SSR markers and their exploitation for assessment of genetic diversity. Biologia 63:227–235

Solin NWNM, Toruan-Mathius N (2014) Genetic diversity of D×P population yield component in oil palm's paternal half-sib family based on microsatellite markers. Energy Procedia 47:196–203

Squirrell J, Hollingsworth P, Woodhead M, Russell J, Lowe A, Gibby M, Powell W (2003) How much effort is required to isolate nuclear microsatellites from plants? Mol Ecol 12:1339–1348

Taeprayoon P, Tanya P, Lee S-H, Srinives P (2015) Genetic background of three commercial oil palm breeding populations in Thailand revealed by SSR markers. Aust J Crop Sci 9:281

Taeprayoon P, Tanya P, Kang YJ, Limsrivilai A, Lee S-H, Srinives P (2016) Genome-wide SSR marker development in oil palm by Illumina HiSeq for parental selection. Plant Genet Resour 14:157–160

Teh C-K et al (2016) Genome-wide association study identifies three key loci for high mesocarp oil content in perennial crop oil palm. Sci Rep 6:19075

Thawaro S, Te-chato S (2010) Verification of legitimate tenera oil palm hybrids using SSR and propagation of hybrids by somatic embryogenesis. Songklanakarin J Sci Technol 32

Ting N-C et al (2010) SSR mining in oil palm EST database: application in oil palm germplasm diversity studies. J Genet 89:135–145
Ting N-C et al (2014) High density SNP and SSR-based genetic maps of two independent oil palm hybrids. BMC Genom 15:309
Tranbarger TJ, Dussert S, Joët T, Argout X, Summo M, Champion A, Cros D, Omore A, Nouy B, Morcillo F (2011) Regulatory mechanisms underlying oil palm fruit mesocarp maturation, ripening, and functional specialization in lipid and carotenoid metabolism. Plant Physiol 156:564–584
Uthaipaisanwong P, Somyong S, Tangphatsornruang S, Yoocha T, Jantasuriyarat C (2017) Development and characterization of simple sequence repeats derived from mitochondrial genome of oil palm using next generation sequencing. Thai J Sci Technol 6:288–300
Varshney R et al (2008) Identification and validation of a core set of informative genic SSR and SNP markers for assaying functional diversity in barley. Mol Breeding 22:1–13
Wang DG et al (1998) Large-scale identification, mapping, and genotyping of single-nucleotide polymorphisms in the human genome. Science 280:1077–1082
Weising K, Nybon H, Wolff K, Kahl G (2005) Applications of DNA fingerprinting in plant sciences. DNA Fingerprinting in plants-principles, methods, and applications, pp 235–276
Welsh J, McClelland M (1990) Fingerprinting genomes using PCR with arbitrary primers. Nucleic Acids Res 18:7213–7218
Xia W et al (2019) Development of high-density SNP markers and their application in evaluating genetic diversity and population structure in *Elaeis guineensis*. Front Plant Sci 10:130
Xiao Y, Zhou L, Xia W, Mason AS, Yang Y, Ma Z, Peng M (2014) Exploiting transcriptome data for the development and characterization of gene-based SSR markers related to cold tolerance in oil palm (*Elaeis guineensis*). BMC Plant Biol 14:384
Xiao Y et al (2016) Genome-wide identification and transferability of microsatellite markers between Palmae species. Front Plant Sci 7:1578
Zaki NM, Ismail I, Rosli R, Chin TN, Singh R (2010) Development and characterization of *Elaeis oleifera* microsatellite markers. Sains Malaysiana 39:909–912
Zaki NM, Singh R, Rosli R, Ismail I (2012) Elaeis oleifera genomic-SSR markers: exploitation in oil palm germplasm diversity and cross-amplification in Arecaceae. Int J Mol Sci 13:4069–4088
Zane L, Bargelloni L, Patarnello T (2002) Strategies for microsatellite isolation: a review. Mol Ecol 11:1–16
Zulkifli Y, Maizura I, Rajinder S (2009) Genetic diversity study of *Elaeis guineensis* germplasm using EST-SSRs. In: Actas de la international society for oil palm breeders (ISOPB) seminar. KLCC, Kuala Lumpur

Genetic Dissecting Complex Traits via Conventional QTL Analysis and Association Mapping

9

Chee-Keng Teh, Nurshazwani Amalina Sudirman, Fairuz Farhana Mohd Rodzik, Ai-Ling Ong, Qi-Bin Kwong, and David Ross Appleton

Abstract

Oil palm yield has been stagnated for decades due to long selection cycle; therefore, marker-assisted selection was initiated. Many quantitative trait loci (QTL) for complex traits were dissected using conventional linkage analyses, but the results were inconsistent and unreproducible across families. The largest challenge is mainly limited recombination in small families leading to detection failure of QTLs. To enhance the detection power, genome-wide association studies (GWAS) access total recombination that accumulated in large multi-parental populations. However, false-positive signals must be monitored and controlled. The recent GWAS for mesocarp oil content revealed the major QTLs on Chromosome 5. This is where the researcher can start mining the candidate genes for further functional studies. In addition, prediction accuracy of genomic selection also can be improved by understanding the QTLs that have been investigated in GWAS. Many genetic tools and knowledge are well established now. This is the time for researchers to pick the right tools and plant materials to address the needs of the oil palm industry.

9.1 Introduction

In 1977, the first DNA-based genome of a ΦX174 bateriophage with a size of approximately 5375 bp was successfully sequenced using Sanger method (Sanger et al. 1977). Since then, enormous efforts have been put into for method improvement and automation, particularly after Human Genome Project (HGP) was established. The Sanger sequencing was then succeeded by the second DNA sequencing revolution in 2005 and now is dominated by New Genome Sequencing (NGS) technology. Sequencing cost and errors have been significantly reduced and the trend persists. In year 2017 alone, number of genome projects in Genome OnLine Database reached 22785 (https://gold.jgi.doe.gov/). The remarkable progress has enabled direct assessment of genetic polymorphisms throughout the genome in nucleus, mitochondria, and chloroplast. This leads to the question, how do the DNA sequencing data benefit us, especially in oil palm?

In the genome, the common types of variation include point mutation, known as single nucleotide polymorphism (SNP), small insertion-and-deletion (indel), and large structural rearrangement of the DNA sequence. These variations are first exploited as DNA markers to

C.-K. Teh (✉) · N. A. Sudirman · F. F. M. Rodzik · A.-L. Ong · Q.-B. Kwong · D. R. Appleton
Biotechnology and Breeding Department, Sime Darby Plantation R&D Centre, 43400 Serdang, Selangor, Malaysia
e-mail: teh.chee.keng@simedarbyplantation.com

© Springer Nature Switzerland AG 2020

M. Ithnin and A. Kushairi (eds.), *The Oil Palm Genome*, Compendium of Plant Genomes, https://doi.org/10.1007/978-3-030-22549-0_9

survey evolution history of a species. Origins of human from Sub-Saharan Africa (Li et al. 2008) and Southeast Asia (The-HUGO-Pan-Asian-SNP-Consortium 2009) were reported. Similarly, germplasm of two oil palm species maintained in Malaysia Palm Oil Board (MPOB) were also studied using simple sequence repeat (SSR) markers, revealing the centers of diversity for *Elaeis guineensis* (Bakoumé et al. 2015) and *E. oleifera* (Maizura et al. 2017) in West Africa and Latin America, respectively. The studies provided an important guideline for both conservation and breeding programs. In fact, the importance of genetic variation becomes more manifest as the inheritance of genic or other portions of the genome are identified to be responsible for phenotypic changes, influencing risk of disease and fitness in the environment. The first application is using the DNA-based diagnosis tools to quantify and predict disease susceptibility and risk at birth. More than 1800 disease-related genes were discovered based on the published genomes under HGP. Now, one of the global largest pharmaceutical firms, AstraZeneca and collaborators, are pushing the boundaries to sequence two million human genomes and associate with clinical trial data to further understand the rare diseases and the response to treatments (Ledford 2016). The success story of human genomics has paved the way for animal and plant breeding. However, the pace is much slower due to resource constraints. In this chapter, the focus will be on genetic dissection of complex traits and milestones achieved in oil palm.

Since the first oil palm plantation was established, the commercial African species (*E. guineensis*) only underwent 5–6 generations of selection (Hartley 1967). Based on monogenic fruit shell trait (see Chap. 4 and this chapter), reciprocal recurrent selection is commonly deployed to select the elite maternal and paternal parents maintained in separate gene pools. Each breeding cycle requires about 12 years to complete and the most efforts is conducting progeny testing. Hence, the breeders are eager to explore possibilities of marker-assisted selection (MAS) to expedite breeding progress with minimal progeny test required. However, MAS is not feasible without knowing the QTL for traits of interest. This chapter will walk through the evolution of quantitative trait loci (QTL) mapping methods used in oil palm and applications. Some challenges and important lessons will also be discussed later.

9.2 Simple and Complex Traits

Like other crops, agronomic traits in oil palm can be either quantitative or qualitative. The simple traits are usually qualitative, following Mendelian inheritance with high penetrance. Usually, these simple traits under control of one or a few genes demonstrate a complete, over or co-dominance effect. For oil palm, the famous simple traits are fruit shell and fruit color, which are controlled by a codominant *SHELL* gene (Singh et al. 2013a) and a complete dominant *VIR* gene (Singh et al. 2014), respectively. The thin-shelled *tenera* ($Sh^{+}Sh^{-}$) can be easily fixed and manipulated just by hybridizing between thick-shelled *dura* ($Sh^{+}Sh^{+}$) and shell-less *pisifera* ($Sh^{-}Sh^{-}$). A typical example, switching from planting *dura* to *tenera* in the 1960s immediately increased 30% oil yield (Corley and Tinker 2003; Hardon et al. 1987). The next major increment has happened after introduction of pollinating weevils (*Elaeidobius kamerunicus*) from West Africa to Malaysia in 1981 and subsequently to Indonesia, Papua New Guinea, and Colombia (Kang 1999; Syed 1982).

Since the 80's, oil yield of commercial plantings has been stagnated at about 4 t/ha/yr. Many oil yield traits are quantitative and sensitive to environmental influence, causing deceleration of selection response. Hence, these traits are also known to be complex. A good example of complex trait is fresh fruit bunch (FFB) that depends on a ratio between male and female inflorescence on a plant. Male inflorescence production is promoted during dry season, leading to lower FFB (Adam et al. 2011). Besides of the yield traits, the industry players are also keen to improve oil quality, *Ganoderma* resistance, dwarfism, abiotic stress resistance, and so on. All these quantitative traits have different degrees of complexity, which can be measured in terms of heritability. This is a statistic used to estimate the phenotypic variation in a population that is due to genetic variation between individuals. Now, the geneticists even can estimate genomic heritability using a linear regression on a set of DNA markers (de los Campos et al. 2015). For oil palm, the fruitlet component traits (composition of mesocarp, shell, and kernel) have higher genomic heritability than that of fruit-to-bunch ratio (F/B) and oil-to-mesocarp ratio (O/M) (Kwong et al. 2017). The poor heritability of F/B as a measurement of fruit formation on a bunch is partly due to strong dependence on weevil population abundance in different climatic conditions (Agus et al. 2014). In many cases, heritability is normally inversely correlated to trait complexity. A reference of heritability is crucial to maximize the chance of QTL detection. Figure 9.1 summarizes the relations among trait complexity, heritability, and practical QTL mapping methods. Bi-parental linkage mapping performs well for simple to moderate traits, while association analysis and genomic selection are more powerful for studying complex traits.

9.3 Conventional Linkage Mapping

The first genetic map based on morphological traits of fruit flies (*Drosophila melanogaster*) was constructed using linkage analysis in year 1913 (Sturtevant 1913), even 40 years earlier than discovery of DNA molecule. The method exploits meiotic recombination between loci (genes or markers) within a family (also known as mapping population). The term linkage refers to the co-inheritance of alleles of loci that are located on the same chromosome from parents to offspring, whereas recombination tends to happen in longer distance between loci. Hence, this explains how linkage distance (in unit centimorgan, cM) between loci is determined by recombination frequency (RF). Each centimorgan equals to 1% RF, but this does not reflect the physical distance between loci. Currently, the same concept is still being used to map DNA markers with the aim of understanding genome architecture of a species, rather than relying on trait-to-trait recombination. This has enabled construction of linkage map as a reference genome to localize QTL or genes responsible for traits based on correlation between genetic polymorphism and phenotypic variation in an unbiased manner.

Indeed, DNA marker deployment in oil palm can be traced back to the early 90s. The first genetic map of oil palm was constructed using codominant restricted fragment length polymorphism (RFLP) markers to identify *Sh* locus for the fruit shell trait (Mayes et al. 1997). With 98 palms derived from a self-pollinated *tenera* family, a total of 97 RFLP markers were linked into 24 linkage groups (LGs) and the map spanned 860 cM in total. The mapping density was then slightly improved to 131 markers based on the first RFLP map by Mayes et al. (1997) to identify QTLs for fruit characteristics and vegetative traits (Rance et al. 2001). High marker density was unachievable at that time because of low throughput of RFLP markers. The throughput was then successfully increased by migrating to PCR-based marker systems, including random amplified polymorphic DNA (RAPD) and amplified fragment length polymorphism (AFLP). In oil palm, linkage map saturation using AFLP and SSR were encouraging (Billotte et al. 2005; Seng et al. 2011; Singh et al. 2009), especially in solving reproducibility problem of RAPD. Small sequence of AFLP markers with 140–350 bp were identified in oil palm (Seng et al. 2007), but the markers cannot directly be used for MAS due to unreliable specificity (Brugmans et al. 2003; Negi et al. 2000). To solve this, the AFLP markers must be converted to other sequence-specific markers, such as sequence characterized amplified region (SCAR) and cleaved amplified polymorphisms (CAPS).

High reproducibility, automation, and affordable cost of the marker systems have then enabled construction of high-density linkage maps which provide a useful mean of identifying QTL for complex traits, including oil yield components (Billotte et al. 2010), palm height (Lee et al. 2015), fatty acid composition (Montoya et al. 2013; Singh et al. 2009), culture

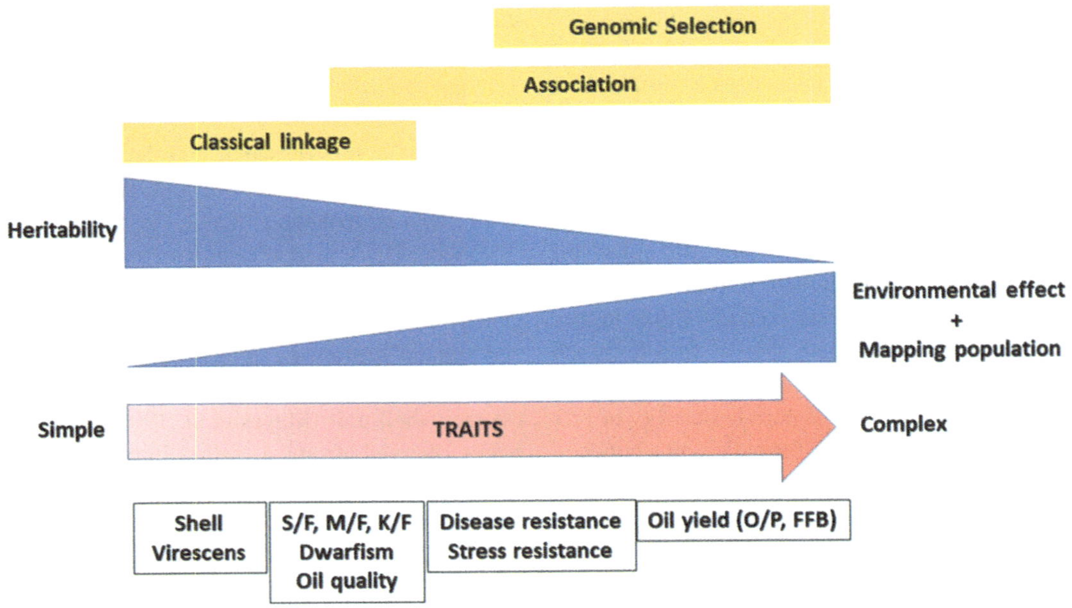

Fig. 9.1 A summary chart of relationship among trait complexity, heritability and practical QTL mapping methods. S/F—Shell-to-fruit ratio; M/F—Mesocarp-to-fruit ratio; K/F—Kernel-to-fruit ratio; O/P—oil per palm; FFB—Fresh fruit bunch

amenability (Ting et al. 2013), and palm oil acidity (Domonhédo et al. 2018). In the past 20 years, oil palm research communities joined other crops like maize, rice, and barley using high-density SNP markers. A 4.5 k custom Illumina SNP array was developed from two oil palm families and the marker density was further improved to 200 k, namely OP200K Illumina array representing 59 important breeding origins (Kwong et al. 2016). Besides genotyping array, a total of 11394 informative SNP markers were also developed in a Deli × AVROS family using a genotyping-by-sequencing method, RAD-seq (Bai et al. 2018). Nevertheless, most of the genetic discoveries in oil palm were the results from adopting the conventional linkage analysis. The family-based approach coupled with high marker density is reasonably powerful for detecting the major QTLs particularly for simple traits. However, inconsistent and ambiguous QTLs for complex traits were observed across origins, which were also observed at the early stage of human genetic mappings. Linkage analysis was reported to be impractical for non-Mendelian diseases in the 80s, mainly due to small family size, non-designed crossing, and non-systematic genetic markers (Altshuler et al. 2008).

The above-mentioned limitations are also common in many perennial crops, like oil palm. Mapping populations were usually designed for conventional breeding, rather than linkage analysis. A single breeding cross typically produces more than 1000 seeds, but only 16–64 palms are field planted and evaluated. Thus, poor recombination event in small populations can limit the genome mappability and QTL detection, especially at centromeric region of chromosomes. Species with larger genome size often indicate higher RF difference between telomeres and centromeric region of chromosomes (Jordan et al. 2018; Sidhu and Gill 2005). One way to determine the linkage coverage of a genome is to run a comparison between linkage position and physical position of pseudomolecules, which also allows determining recombination landscape of whole genome (Blair et al. 2018; Teh et al. 2017b). For a good genome coverage, all chromosome-linkage group comparison should fit with sigmoidal curves, with steep lines of high recombination of both ends and flatter plateaus of low recombination at the middle (Fig. 9.2).

Thus, oil palm researchers put up enormous effort to saturate genetic map by exhaustedly include all informative markers identified in single-cross populations. In the last five years, linkage map of oil palm with 1.40 cM mapping interval (Ting et al. 2014) was improved to 1.26 cM (Pootakham et al. 2015) and followed by the highest interval, 0.29 cM (Bai et al. 2018) as reported to date. Without expanding population size, the saturation on centromeric regions can be marginal because of stagnated RF in the populations. Another problem arises with too many linked markers can substantially increase the significant threshold of LOD score determined from permutation test (Zeng 1994). Naive threshold of 3.0 LOD score is usually set to salvage any potential QTLs, but this can be prone to false-positive results as well. To avoid this, identical and closely linked markers can be removed to generate evenly spaced map with ≤ 10 cM mapping density to ensure low mapping error caused by double recombination (Xu et al. 2005). For oil palm, the key challenge is to increase the family size, targeting for higher recombination rate. One practical solution is utilizing all the genetic polymorphisms present in the existing breeding set up. The conventional breeding normally evaluates many families, sometimes from different origins. In this case, hundreds or thousands of offspring created for progeny test are accessible and confer a good basis for association mapping and genomic selection.

9.4 Association Mapping and Genomic Selection

Association mapping can be implemented at the level of candidate genes and whole genome. The candidate gene approach has been the leading genetic association studies, especially in identifying risk variants associated with a disease for

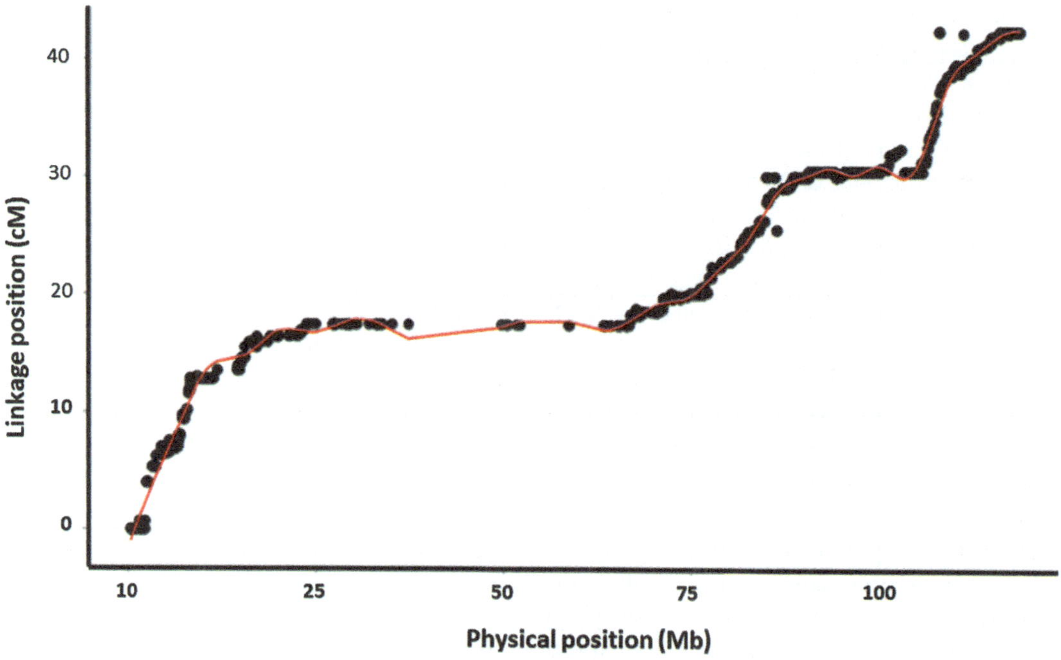

Fig. 9.2 A correlation plot between linkage position (in centimorgan) and physical position of a pseudomolecule (in Mb). Sigmoidal curve indicates coverage of linkage and physical map from a telomere to another telomere, and the plateau in the middle is centromeric region

human. The approach is cheap and easy to perform, but it requires prior knowledge of putative candidate genes and their relevance in the mechanism of the trait being investigated (Kwon and Goate 2000). The approach was usually challenged for non-reproducible (Patnala et al. 2013) and false-negative findings (type II error) (Braem et al. 2011; Pharoah et al. 2004). The latter can be the results from lack of statistical power of small sample size and under evaluation of the gene–gene and gene–environment interactions (Pharoah et al. 2004). Again, this explains the need of having large sample size to improve the accuracy of QTL mapping. In some cases, penetrance of causal variants for complex traits can be too small and hard to detect through co-segregation within families or pedigrees. Therefore, the whole-genome association, namely GWAS was first introduced in human since year 2002. The method is based upon the principle of linkage disequilibrium (LD) in large populations. Unlike linkage analysis, GWAS utilize the total recombination due to evolutionary forces, such as mutation, genetic drift, and selection that have been accumulated in a population (Hartl and Clark 1997). This confers higher mapping resolution on GWAS by breaking down the long LDs through the abundant recombination.

The publication of the oil palm genome consisting of 1.535 Gb of assembled sequenced (Singh et al. 2013b), the independent assembly of Sime Darby Plantation's oil palm genome (unpublished) and high-density OP200K SNP array (Kwong et al. 2016), has enabled the first QTL discovery for mesocarp oil content of oil palm via GWAS (Teh et al. 2016). The genomic distance of LD decays determined the number of SNP markers required to tag a least one haplotype throughout the genome. As expected, the commercial Deli × AVROS population had longer LD decay (25 Kb at 0.12 of average pairwise correlation coefficient (r^2)) than the semi-wild Deli × Nigerian population (20 Kb at 0.15 of r^2). These observations also showed that the OP200K array with average sampling of a SNP marker for every 11 Kb based on the reference genome size, provided sufficient marker density for GWAS. The genome-wide approach is known to be susceptible to confounding factor, especially if population structure and cryptic relatedness are present in the assayed population (Astle and Balding 2009). The factor here is referring to the inflated false-positive association signals which explain the variance between substructures, instead of targeted phenotypic variance. Many correction methods, such as genomic control, Bonferroni correction, and false discovery rate control, have been introduced to address the false positives. The applicability of each correction method is on case-by-case basis. For example, corrections like Bonferroni method can be too stringent causing over-correction. Genomic inflation factor (GIF) and quantile-quantile plot of -$\log_{10}$ p-value can be adopted to monitor the false-positive SNPs and to decide the best correction method. The optimal result with significant QTL detection will be GIF close to 1.0 under the null hypothesis and deviation from null distribution at the tail of a quantile-quantile plot. After the corrections, the major QTLs for mesocarp oil content (O/DM) in both commercial and semi-wild oil palm populations are mainly situated on Chromosome 5 (Teh et al. 2016).

Eventually, GWAS can provide a list of candidate genes depending on the length of LD blocks for further functional studies to understand the underlying causal variants responsible for traits of interest. Indeed, GWAS result alone can be directly used to initiate MAS program in oil palm, even without knowing the candidate genes involved. Conventional MAS only pick a few major QTLs for breeding prediction. The genetic effects of the selected QTLs for complex trait can be very small; therefore, selection response becomes insignificant (Bernardo 2008). To avoid this, effect of all markers throughout the genome is consolidated to estimate total breeding value, which is genomic selection (Meuwissen et al. 2001). Many studies were conducted for oil palm in recent years and their prediction abilities were promising (Cros et al. 2015; Kwong et al. 2017; Wong and Bernardo 2008). Unlike the conventional MAS, genomic selection does not need any information on physical position of assayed markers and QTL. Interestingly, prediction accuracy was found to be higher when genomic selection was guided by GWAS results (Kwong et al. 2017).

In 2016, Sime Darby Plantation began commercializing genomic tools through the development and the first full-scale planting of GenomeSelect™ seeds that were genetically selected for high yield traits. The heritability (h^2) in 'The Breeder's Equation' ($R = h^2S$) is substituted by prediction accuracy of genomic selection to estimate the selection response (R_a) within a *tenera* generation (Teh et al. 2017a). However, genotyping every *tenera* seed in a full-scale seed production is tedious and costly, mainly due to large sample size and marker density. An established producer normally generates more than a million seeds per year; thus, it requires high capacity for genotyping and selection of top-ranking individuals. One method used to address this was the fine-tuning of genomic selection model and their application on mother and pollen palms, which only require one round of genotyping. Mass production of GenomeSelect™ seeds is now directly being derived from the selected *dura* × *pisifera* combinations. Marker density also can be reduced based on the LD filtering approach without compromising prediction accuracy (Kwong et al. 2017). By doing both, more than 95% genotyping cost was successfully reduced and enabled increased production scale of GenomeSelect™ seeds for 1000 ha planting per year (Fig. 9.3). The ultimate aim is to achieve annual 5% replanting rate of Sime Darby Plantation by year 2021 onwards.

9.5 Conclusion

The rapid development in DNA sequencing and molecular technologies for the past decades has revolutionized the genetic discoveries in oil palm. Today, informative DNA markers are readily generated and no longer a limiting factor. Understanding the trait complexity and heritability is the key success factor to ensure optimal

Fig. 9.3 Part of GenomeSelect™ 500 ha planting at Sungai Buloh Estate

experimental designs such as crossing programs and sample size. Concurrent advancement of bioinformatics tool also allows deep quantitative genetic analysis of large dataset. In other words, oil palm research is now at a good position to dissect the complex traits.

Acknowledgements We would like to express our appreciation to the book editor, Dr. Maizura Ithnin for inviting us. We also thank Sime Darby Plantation for supporting all the research works.

References

Adam H, Collin M, Richaud F, Beulé T, Cros D, Omoré A, Nodichao L, Nouy B, Tregear JW (2011) Environmental regulation of sex determination in oil palm: current knowledge and insights from other species. Ann Bot 108:1529–1537

Agus EP, Wiharti OP, Agus S (2014) *Elaeidobius kamerunicus*: application of hatch and carry technique for increaseing oil palm fruit set. J Oil Palm Res 26:195–202

Altshuler D, Daly MJ, Lander ES (2008) Genetic mapping in human disease. Science (New York, NY) 322:881–888

Astle W, Balding DJ (2009) Population structure and cryptic relatedness in genetic association studies. Statist Sci 24:451–471

Bai B, Wang L, Zhang YJ, Lee M, Rahmadsyah R, Alfiko Y, Ye BQ, Purwantomo S, Suwanto A, Chua N-H, Yue GH (2018) Developing genome-wide SNPs and constructing an ultrahigh-density linkage map in oil palm. Sci Rep 8:691

Bakoumé C, Wickneswari R, Siju S, Rajanaidu N, Kushairi A, Billotte N (2015) Genetic diversity of the world's largest oil palm (*Elaeis guineensis* Jacq.) field genebank accessions using microsatellite markers. Genet Resour Crop Evol 62:349–360

Bernardo R (2008) Molecular markers and selection for complex traits in plants: learning from the last 20 years. Crop Sci 48:1649–1664

Billotte N, Marseillac N, Risterucci AM, Adon B, Brottier P, Baurens FC, Singh R, Herrán A, Asmady H, Billot C, Amblard P, Durand-Gasselin T, Courtois B, Asmono D, Cheah SC, Rohde W, Ritter E, Charrier A (2005) Microsatellite-based high density linkage map in oil palm (*Elaeis guineensis* Jacq.). Theor Appl Genet 110:754–765

Billotte N, Jourjon MF, Marseillac N, Berger A, Flori A, Asmady H, Adon B, Singh R, Nouy B, Potier F, Cheah SC, Rohde W, Ritter E, Courtois B, Charrier A, Mangin B (2010) QTL detection by multi-parent linkage mapping in oil palm (*Elaeis guineensis* Jacq.). Theor Appl Genet 120:1673–1687

Blair MW, Cortés AJ, Farmer AD, Huang W, Ambachew D, Penmetsa RV, Carrasquilla-Garcia N, Assefa T, Cannon SB (2018) Uneven recombination rate and linkage disequilibrium across a reference SNP map for common bean (*Phaseolus vulgaris* L.). PLOS ONE 13:e0189597

Braem MGM, Schouten LJ, Peeters PHM, den Brandt PAV, Onland-Moret NC (2011) Genetic susceptibility to sporadic ovarian cancer: a systematic review. Biochimica et Biophysica Acta (BBA)—Rev Cancer 1816:132–146

Brugmans B, van der Hulst RGM, Visser RGF, Lindhout P, van Eck HJ (2003) A new and versatile method for the successful conversion of AFLP™ markers into simple single locus markers. Nucleic Acids Res 31:e55

Corley RHV, Tinker PB (2003) Selection and breeding. The oil palm, 3ed edn. Blackwell, pp 133–187

Cros D, Denis M, Sánchez L, Cochard B, Flori A, Durand-Gasselin T, Nouy B, Omoré A, Pomiès V, Riou V, Suryana E, Bouvet J-M (2015) Genomic selection prediction accuracy in a perennial crop: case study of oil palm (*Elaeis guineensis* Jacq.). Theor Appl Genet 128:397–410

de los Campos G, Sorensen D, Gianola D (2015) Genomic heritability: what is it? PLOS Genet 11: e1005048

Domonhédo H, Cuéllar T, Espeout S, Droc G, Summo M, Rivallan R, Cros D, Nouy B, Omoré A, Nodichao L, Arondel V, Ahanhanzo C, Billotte N (2018) Genomic structure, QTL mapping, and molecular markers of lipase genes responsible for palm oil acidity in the oil palm (*Elaeis guineensis* Jacq.). Tree Genet Genomes 14:69

Hardon JJ, Corley RHV, Lee CH (1987) Breeding and selecting the oil palm. Academic Press, London

Hartl DL, Clark AG (1997) Principles of population genetics. Sinauer Associates, Sunderland

Hartley CWS (1967) The origin and development of the oil palm industry, The Oil Palm, 1st edn. Longman, London, pp 1–36

Jordan KW, Wang S, He F, Chao S, Lun Y, Paux E, Sourdille P, Sherman J, Akhunova A, Blake NK, Pumphrey MO, Glover K, Dubcovsky J, Talbert L, Akhunov ED (2018) The genetic architecture of genome-wide recombination rate variation in allopolyploid wheat revealed by nested association mapping. Plant J Cell Mol Biol 95:1039–1054

Kang SM (1999) The *Elaeidobius kamerunicus* story. Kuala Lumpur

Kwon JM, Goate AM (2000) The candidate gene approach. Alcohol Res Health 24:164–168

Kwong QB, Teh CK, Ong AL, Heng HY, Lee HL, Mohamed M, Low JB, Apparow S, Chew FT, Mayes S, Kulaveerasingam H, Tammi M, Appleton DR (2016) Development and validation of a high-density SNP genotyping array for African oil palm. Mol Plant 9:1132–1141

Kwong QB, Ong AL, Teh CK, Chew FT, Tammi M, Mayes S, Kulaveerasingam H, Yeoh SH,

Harikrishna JA, Appleton DR (2017) Genomic selection in commercial perennial crops: applicability and improvement in oil palm (*Elaeis guineensis* Jacq.). Sci Rep 7:2872

Ledford H (2016) Drug firm seeks genome bounty. Nature. https://www.nature.com/polopoly_fs/1.19797!/menu/main/topColumns/topLeftColumn/pdf/nature.2016.19797.pdf?origin=ppub

Lee M, Xia JH, Zou Z, Ye J, Rahmadsyah Alfiko Y, Jin J, Lieando JV, Purnamasari MI, Lim CH, Suwanto A, Wong L, Chua N-H, Yue GH (2015) A consensus linkage map of oil palm and a major QTL for stem height. Sci Rep 5:8232

Li JZ, Absher DM, Tang H, Southwick AM, Casto AM, Ramachandran S, Cann HM, Barsh GS, Feldman M, Cavalli-Sforza LL, Myers RM (2008) Worldwide human relationships inferred from genome-wide patterns of variation. Science 319:1100

Maizura I, Teh C-K, Wickneswari R (2017) Genetic diversity of *Elaeis oleifera* (HBK) Cortes populations using cross species SSRs: implication's for germplasm utilization and conservation. BMC Genet 18:37

Mayes S, Jack PL, Corley RHV, Marshall DF (1997) Construction of a RFLP genetic linkage map for oil palm (*Elaeis guineensis* Jacq.). Genome 40:116–122

Meuwissen THE, Hayes BJ, Goddard ME (2001) Prediction of total genetic value using genome-wide dense marker maps. Genetics 157:1819–1829

Montoya C, Lopes R, Flori A, Cros D, Cuellar T, Summo M, Espeout S, Rivallan R, Risterucci AM, Bittencourt D, Zambrano JR, Alarcón GWH, Villeneuve P, Pina M, Nouy B, Amblard P, Ritter E, Leroy T, Billotte N (2013) Quantitative trait loci (QTLs) analysis of palm oil fatty acid composition in an interspecific pseudo-backcross from *Elaeis oleifera* (H.B.K.) Cortés and oil palm (*Elaeis guineensis* Jacq.). Tree Genet Genomes 9:1207–1225

Negi MS, Devic M, Delseny M, Lakshmikumaran M (2000) Identification of AFLP fragments linked to seed coat colour in *Brassica juncea* and conversion to a SCAR marker for rapid selection. Theor Appl Genet 101:146–152

Patnala R, Clements J, Batra J (2013) Candidate gene association studies: a comprehensive guide to useful in silicotools. BMC Genet 14:39

Pharoah PDP, Dunning AM, Ponder BAJ, Easton DF (2004) Association studies for finding cancer-susceptibility genetic variants. Nat Rev Cancer 4:850

Pootakham W, Jomchai N, Ruang-areerate P, Shearman JR, Sonthirod C, Sangsrakru D, Tragoonrung S, Tangphatsornruang S (2015) Genome-wide SNP discovery and identification of QTL associated with agronomic traits in oil palm using genotyping-by-sequencing (GBS). Genomics 105:288–295

Rance KA, Mayes S, Price Z, Jack PL, Corley RHV (2001) Quantitative trait loci for yield components in oil palm (*Elaeis guineensis* Jacq.). Theor Appl Genet 103:1302–1310

Sanger F, Air GM, Barrell BG, Brown NL, Coulson AR, Fiddes JC, Hutchison Iii CA, Slocombe PM, Smith M (1977) Nucleotide sequence of bacteriophage ΦX174 DNA. Nature 265:687

Seng TY, Faridah QZ, Ho CL, Maizura I, Rao V (2007) Flanking AFLP markers for the Virescens trait in oil palm. J Oil Palm Res 19:381–392

Seng T-Y, Mohamed Saad SH, Chin C-W, Ting N-C, Harminder Singh RS, Qamaruz Zaman F, Tan S-G, Syed Alwee SSR (2011) Genetic linkage map of a high yielding FELDA Deli × Yangambi oil palm cross. PLoS ONE 6:e26593

Sidhu D, Gill KS (2005) Distribution of genes and recombination in wheat and other eukaryotes. Plant Cell Tissue Organ Cult 79:257–270

Singh R, Tan SG, Panandam JM, Rahman RA, Ooi LCL, Low E-TL, Sharma M, Jansen J, Cheah S-C (2009) Mapping quantitative trait loci (QTLs) for fatty acid composition in an interspecific cross of oil palm. BMC Plant Biol 9:114–114

Singh R, Low E-TL, Ooi LC-L, Ong-Abdullah M, Ting N-C, Nagappan J, Nookiah R, Amiruddin MD, Rosli R, Manaf MAA, Chan K-L, Halim MA, Azizi N, Lakey N, Smith SW, Budiman MA, Hogan M, Bacher B, Van Brunt A, Wang C, Ordway JM, Sambanthamurthi R, Martienssen RA (2013a) The oil palm SHELL gene controls oil yield and encodes a homologue of SEEDSTICK. Nature 500:340–344

Singh R, Ong-Abdullah M, Low ETL, Manaf MAA, Rosli R, Nookiah R, Ooi LCL, Ooi SE, Chan KL, Halim MA, Azizi N, Nagappan J, Bacher B, Lakey N, Smith SW, He D, Hogan M, Budiman MA, Lee EK, DeSalle R, Kudrna D, Goicoechea JL, Wing RA, Wilson RK, Fulton RS, Ordway JM, Martienssen RA, Sambanthamurthi R (2013b) Oil palm genome sequence reveals divergence of interfertile species in Old and New worlds. Nature 500:335–339

Singh R, Low ETL, Ooi LCL, Ong-Abdullah M, Nookiah R, Ting NC, Marjuni M, Chan PL, Ithnin M, Manaf MAA, Nagappan J, Chan KL, Rosli R, Halim MA, Azizi N, Budiman MA, Lakey N, Bacher B, Van Brunt A, Wang C, Hogan M, He D, MacDonald JD, Smith SW, Ordway JM, Martienssen RA, Sambanthamurthi R (2014) The oil palm VIRESCENS gene controls fruit colour and encodes a R2R3-MYB. Nat Commun 5

Sturtevant AH (1913) The linear arrangement of six sex-linked factors in *Drosophila*, as shown by their mode of association. J Exp Zool 14:43–59

Syed RA (1982) Insect pollination of oil palm: feasibility of introducing *Elaeidobius* spp. into Malaysia, In: Pushparajah E, Chew PS (eds) The oil palm in agriculture in the eighties. Incorp. Soc. Planters, Kuala Lumpur, pp 263–290

Teh C-K, Ong A-L, Kwong Q-B, Apparow S, Chew F-T, Mayes S, Mohamed M, Appleton D, Kulaveerasingam H (2016) Genome-wide association study identifies three key loci for high mesocarp oil content in perennial crop oil palm. Sci Rep 6:19075

Teh CK, Kwong QB, Ong AL, Mohaimi M, Sukganah A, Chew FT, Mayes S, Appleton D, Harikrishna K

(2017a) Application of genomic tools in oil palm breeding. In: Soh AC, Mayes S, Roberts JA (eds) Oil palm breeding: genetics and genomics. CRC Press, Boca Raton, pp 246–256

Teh CK, Ong AL, Kwong QB (2017b) Applications of NGS data. In: Low L, Tammi M (eds) Bioinformatics: a practical handbook of next generation sequencing and its applications. World Scientific, Singapore, pp 195–229

The-HUGO-Pan-Asian-SNP-Consortium (2009) Mapping human genetic diversity in Asia, Science, p 1541

Ting N-C, Jansen J, Nagappan J, Ishak Z, Chin C-W, Tan S-G, Cheah S-C, Singh R (2013) Identification of QTLs associated with callogenesis and embryogenesis in oil palm using genetic linkage maps improved with SSR markers. PLoS ONE 8:e53076

Ting NC, Jansen J, Mayes S, Massawe F, Sambanthamurthi R, Ooi LCL, Chin CW, Arulandoo X, Seng TY, Alwee SSRS, Ithnin M, Singh R (2014) High density SNP and SSR-based genetic maps of two independent oil palm hybrids. BMC Genom 15:309

Wong CK, Bernardo R (2008) Genomewide selection in oil palm: increasing selection gain per unit time and cost with small populations. Theor Appl Genet 116

Xu Z, Zou F, Vision TJ (2005) Improving quantitative trait loci mapping resolution in experimental crosses by the use of genotypically selected samples. Genetics 170:401–408

Zeng ZB (1994) Precision mapping of quantitative trait loci. Genetics 136:1457–1468

Omics—A Potential Tool for Oil Palm Improvement and Productivity

10

Umi Salamah Ramli, Abrizah Othman, Noor Idayu Mhd Tahir, Benjamin Yii Chung Lau, Syahanim Shahwan, Hasliza Hassan, Zain Nurazah, Shahirah Balqis Dzulkafli, Nurul Liyana Rozali, Nur Ain Ishak, Ravigadevi Sambanthamurthi, Ghulam Kadir Ahmad Parveez, and Ahmad Kushairi

Abstract

Palm oil is one of the major sources of edible oils and oleochemical feedstocks. Sustainable oil palm cultivation which is set to boost production requires effective and innovative strategies. There is a continuous effort to increase the yield and productivity of the oil palm through conventional breeding and by cloning super planting materials. Hitherto, oil palm breeders had limited choice of tools to evaluate the important traits of the palm such as resistance to diseases which has constrained the breeding programme. Genomics-based technologies have sped up the process. Post-genomics tools such as transcriptomics, proteomics and metabolomics are well-established technologies and have been used as phenotyping tools to elucidate the mechanisms involved in fruit ripening and fatty acid synthesis, all of which promise to facilitate and speed up the pace of oil palm improvement. Oil palm diseases also have major economic repercussions for the oil palm industry. Progress in omics studies aimed to advance the knowledge in plant-pathogen interactions is discussed, and the process of discovering novel biomarkers and potential therapeutic targets may be shortened using proteomic and metabolomic approaches. Information and the discoveries from these studies have opened the door for the development of an oil palm omics database, which gathers proteome and metabolome data for studies of oil palm systems biology.

U. S. Ramli (✉) · A. Othman · N. I. M. Tahir · B. Y. C. Lau · S. Shahwan · H. Hassan · Z. Nurazah · S. B. Dzulkafli · N. L. Rozali · N. A. Ishak · R. Sambanthamurthi · G. K. A. Parveez · A. Kushairi
Malaysian Palm Oil Board, No. 6, Persiaran Institusi, Bandar Baru Bangi, 43000 Kajang, Selangor, Malaysia
e-mail: umi@mpob.gov.my

10.1 Introduction

Demand for edible oils for the food and non-food uses is forecasted to increase to approximately 240 Mt in 2050 in tandem with the rise of human population and consumption (Corley 2009). Palm oil is one of the 17 major edible oils and fats produced globally with Indonesia and Malaysia accounting for 84% of the world's total production in 2016 (Oil World Annual 2017). Among the vegetable oil producers, oil palm (*Elaeis guineensis*) is the highest producing oil crop in the world and roughly accounts for 36% of the vegetable oil supply (OECD/FAO Agricultural Outlook 2017; Woittiez et al. 2017). Oil palm yields two distinct oils—palm oil and palm kernel oil. The different fatty acid profiles of palm oil and palm kernel oil make the oil palm one of the most versatile crops for food and

© Springer Nature Switzerland AG 2020

M. Ithnin and A. Kushairi (eds.), *The Oil Palm Genome*, Compendium of Plant Genomes, https://doi.org/10.1007/978-3-030-22549-0_10

oleochemical applications (Barcelos et al. 2015). Palm oil is rich in carotenoids which confer its deep red colour, and other major components including the semi-saturated and saturated-fatty acids. Palm oil contains oleic and palmitic acids which, respectively, account for 39 and 44% of its total fatty acid composition. Hence, it is a viscous semi-solid oil, even at tropical ambient temperature, and a solid fat in temperate climates (Sundram et al. 2003). To remain sustainable, the oil palm industry is constantly under global pressure to increase its productivity rather than cultivated area (Chong et al. 2017). Other challenges include the increase in labour cost, declining available arable land for commercial cultivation and diseases that have been seriously affecting overall yield per hectare (Murphy 2014). Thus, concerted actions have to be taken to triumph over these challenges.

The oil palm is a diploid ($2n = 32$) with an approximate genome size of ~1.8 gigabases (Singh et al. 2013a). The use of DNA-based markers was the first foundation for application of markers in oil palm breeding programmes (Billotte et al. 2005; Mayes et al. 1996, 2000; Cheah et al. 1993) while the more recent advent of genomic sequencing and epigenomics-based technologies have provided powerful means and unprecedented opportunity to improve oil palm breeding efficiency (Ong-Abdullah et al. 2015; Singh et al. 2013a; Singh et al. 2013b). The development of fundamentals in gene cloning and transformation technology for genetic improvement of oil palm is expected to speed up and improve the precision of conventional breeding targeting for oil quality (Parveez et al. 2015). Advantageous outcomes from the development of analytical techniques and data interpretation in post-genomics studies have opened up new possibilities and provided powerful phenotyping tools for unravelling biological mechanisms and identifying predictive biomarkers crucial for metabolic characteristics. A landmark in omics research in oil palm was the development of transcriptomics, proteomics and metabolomics research capacity (Ramli et al. 2016). In recent years, these omics have revolutionized oil palm research. The advent of each stratum has been integrated across each multiple omics to studies of genes (genomics), transcripts (transcriptomics), proteins (proteomics), metabolites (metabolomics), and ultimately the observable characteristics or traits (phenomics) of oil palm. Proteome- and metabolome-based markers, proteins and metabolites are the closest features to phenotypic characterization (Wienkoop et al. 2008).

As illustrated in Fig. 10.1, along with the establishment of a database platform, the streamlined workflow for oil palm proteomics and metabolomics is expected to generate an extensive spectrum for a holistic understanding of oil palm systems biology. This information will help to distinguish and further apply suitable phenotyping tools towards horticultural trait

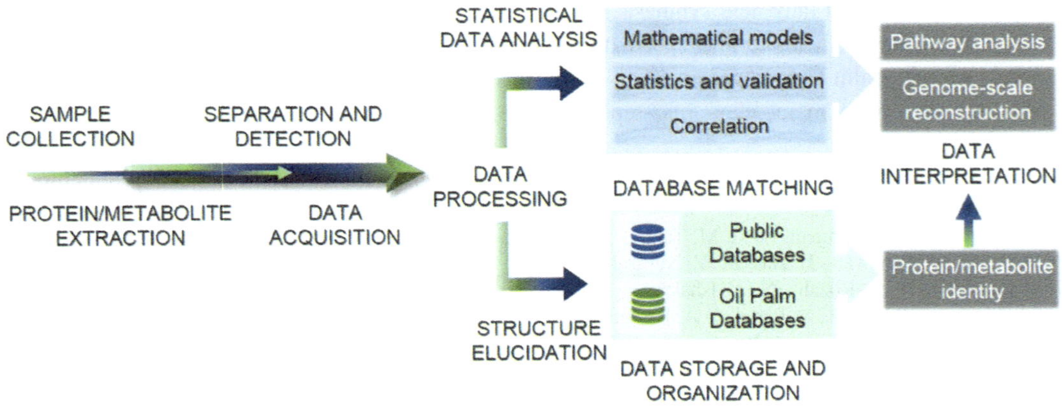

Fig. 10.1 Oil palm proteomics and metabolomics workflow

improvement such as oil quality, yield and diseases. For instance, proteomics and metabolomics platforms are being explored towards understanding the well-coordinated biological processes involved in palm oil production by analysing the fruits harvested at different stages of maturation. Identification of specific proteins and/or metabolic changes associated with infection by *Ganoderma boninense* which causes basal stem rot promises great potential to facilitate development of markers and selection tools for improving productivity and conventional breeding programme for disease tolerance.

10.2 Confronting Challenges in Integrative Approaches for Oil Palm Improvement

Historically, oil palm is propagated by seeds since its commercial planting in the 1920s and has been continuously improved by means of selective breeding (Mayes et al. 2000). Crossing for improved progenies has been facilitated by the extensive oil palm germplasm collection at MPOB and has provided the industry with elite planting materials (Soh et al. 2017). A hundred years have now passed since the first oil palm plantation was established in Malaysia in 1917, and the demand for higher yield is greater than ever. But for a perennial species such as oil palm, the process is long and arduous with its long breeding cycles. Thus, a definitive method of selection to shorten the process and accelerate the breeding programmes is critical. With the availability of a genetic map, utilization of molecular markers in plant breeding has been viewed as a promising approach for large-scale screening for selected traits and improving breeding efficiency. With respect to oil palm, application of markers was initiated in 1990s (Mayes et al. 1996; Cheah et al. 1993). DNA-based markers such as restriction fragment length polymorphism (RFLP), PCR-based random amplified polymorphic (RAPD), amplified fragment length polymorphism (AFLP) and simple sequence repeat (SSR) for genetic mapping introduced new opportunities leading to an effective improvement for breeding and selection (Billotte et al. 2005; Jaligot et al. 2002; Singh and Cheah 1999; Mayes et al. 1997). High-density molecular markers can be used to delineate agronomically desirable traits and to identify candidate genes within a region of interest.

Elucidating phenotypic from genotypic variation is not straightforward (Keurentjes 2009). Thus, investigation of a genome-wide set of genetic variants such as single-nucleotide polymorphisms (SNPs) to discover the association with a specific phenotypic trait, also known as genome-wide association study (GWAS), has been paired with omics techniques such as proteomics and metabolomics for a conclusive correlation. GWAS has provided voluminous amount of data (dubbed as 'big data'), but challenges are acknowledged for complex biological systems whereby in which there are a large number of identified loci, each contributing to a small fraction of the genetic trait. Many-to-one, one-to-many and many-to-many relationships among genomic sequence variants and those at transcript and protein levels bring about inconsistencies in addition to challenges of gene-gene interaction involvement, epistasis and ecological factors (Zhou et al. 2015; Chen and Snyder 2013). All the copious data from gene functional studies emphasized at the gene expression level (transcriptomics) will not manifest into useful information without translation (proteomics) and the biochemical study of metabolites that define the phenotype (Goodacre et al. 2004). This has led to the establishment of post-genomics investigations and robust data mining tools. Precision proteomics, a term coined by Mann and Kelleher (2008), offers large-scale protein analysis using mass spectrometry of high resolving power to elucidate post-translational modifications and polymorphisms. Erstwhile methods of expressing and immobilizing proteins that are low throughput are arduous and have now been replaced with advanced and accurate mass spectrometry analyses. For metabolomic analysis, the types of samples that can be analysed are inclusive, e.g. tissues, cells, in liquid and gaseous forms. Metabolomics approaches provide a powerful strategy for studying localized and specific responses to stimuli and pathogenesis, towards understanding biochemical information underlying the mechanisms of the events.

The prolonged stagnation in the national oil yield at 3.9 tonne per hectare per year (t/ha/yr) for the past two decades impedes growth and productivity of the oil palm industry (Ting et al. 2014). This together with various other factors such as limited land to expand has compelled the R&D sector to focus its research into better oil palm planting materials. Cloning superior palms such as those with high yield would be a way forward and facilitate the achievement of the targeted 26.2 t/ha/yr of fresh fruit bunch by the year 2020, thus providing the industry with a much needed boost (MPOB 2011). However, the aspiration of large-scale elite oil palm clonal propagation is hampered by epigenetic challenges which can be defined as heritable alterations in gene function even without change in the DNA sequence (Bird 2002) so abnormal palms are produced. Epigenetic regulation includes DNA methylation, DNA hydroxymethylation, histone variants and modifications, nucleosome remodelling, and small and large non-coding regulatory RNAs (Han and Garcia 2013). The epigenetic regulation has been mapped on a genomic scale first using techniques based on digestion with restriction enzymes, affinity enrichment (including immune precipitation) or chemical conversion. These techniques are later improved with chemical crosslinking, isotope labelling and affinity purification to minimize false positives (Han and Garcia 2013). They are combined with tagging methods such as isotope and isobaric labels to achieve quantitative proteomics of intact proteins using mass spectrometry to examine the modification sites.

Tissue culture technique is an important tool in complementing an oil palm breeding programme. A reliable protocol and stable system are required to enable mass propagation of elite ortets selected from established D×P progeny trials and also for propagating transgenic oil palm for further breeding and production of improved planting materials. The minimum standard required for the selection of an ortet is 50 kg/palm/year of oil yield and 27% of oil to bunch ratio (MS2099 2008). It can also provide the target tissues, such as callus for transformation of the individual oil palm traits. When ready, genetically modified (GM) oil palm has to be grown in containment for evaluation before it is considered safe to be released to the environment. Although agronomic treatments such as fertilizer application, watering and pruning are standardized across containments and planting fields, some forms of validation are required to investigate the effects of environmental factors such as light penetration and exposure to other biotics towards the plants. Establishment of information on environmental effects such as planting conditions using clones will serve as baseline data to rule out unintended effects of genetic engineering. The metabolome is affected by both genetic and ecological changes (Fiocchi 2014; Fiehn 2002). Plant metabolite regulation thus cannot be deduced simply from genetic data, unlike the transcripts and proteins. The relationship between metabolites and molecular functioning is vague and comparisons need to be made between gene expression QTLs (eQTLS) of biosynthetic genes and metabolite QTLs (Keurentjes 2009). The capturing and mining of the integrated omics data will then require exactness and accuracy of expression or detail to ensure comprehensive interpretation in the field. Figure 10.2 describes the precision omics strategies towards improved crop productivity.

10.3 Exploring Proteomics and Metabolomics in Oil Palm Physiology and Agronomy

Field omics is a comprehensive analysis of large-scale molecular data obtained in field trials. It remains challenging to combine functional genomics to the long-established practice in a crop science which requires field data and careful integration with '-omics' data (Alexandersson et al. 2014). The application of systems-biology-based-approach such as proteomics and metabolomics has the advantages to discover key components in agronomically important traits.

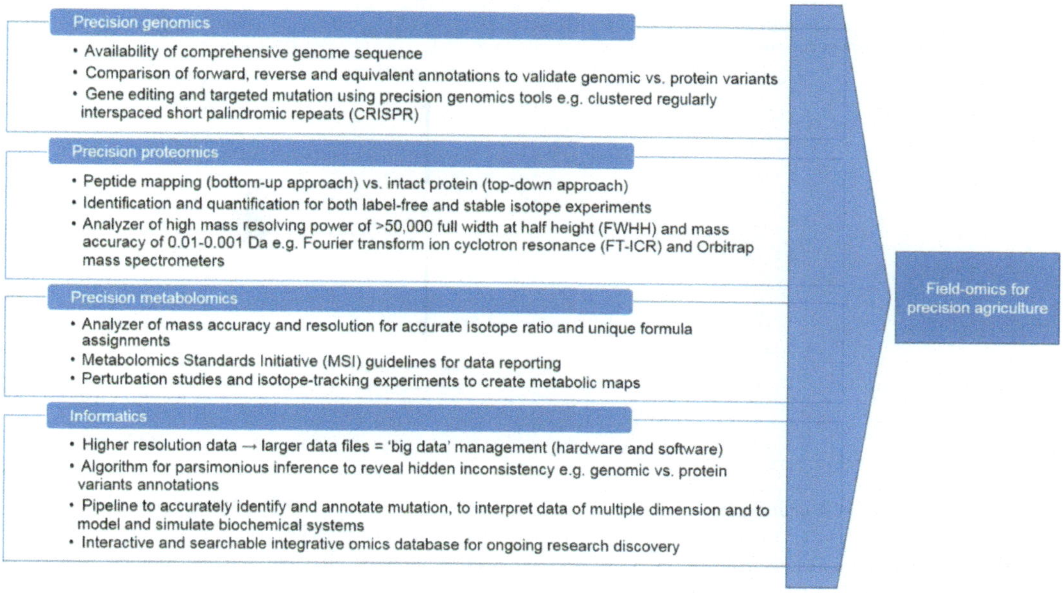

Fig. 10.2 Precision omics strategies towards improved crop productivity

Combining crop physiology and functional genomics data could explain the magnitude of genetic versus environmental perturbations to study factors that drive change of metabolites which in turns determine yield and quality (Alm 2007; Weckwerth 2011; Alexandersson et al. 2014). Oil palm yield is highly influenced by environmental factors such as the soil, agronomic practices and rainfall (Rajanaidu et al. 2000). Characterization of oil palm proteome and metabolome data has profound implications for uncovering the interaction between varying environmental responses to the agronomical traits such as yield and disease resistance. Thus, elucidating complex biological mechanism involved in the interactions between genetic and environmental factors must be pursued diligently using integrated omics approaches incorporating sensitive and robust technologies such as mass spectrometry (Sawada and Hirai 2013; Tohge and Fernie 2015).

A complete understanding of the metabolic pathways requires a combination of experiments that evaluate biochemical reactions including flux in the pathways under specific conditions. Proteomics and metabolomics can play a pivotal role in capturing changes and molecular responses to a variety of environmental stress factors such as planting sites and their biotic and abiotic properties. Metabolomics has been successfully used to evaluate differences at the level of metabolites between clonal palms planted in peat and mineral soils (Tahir et al. 2016). Figure 10.3 shows the scores and loading plots of principal component analysis (PCA) that reveal the abundance of asparagine and dopamine contributing to the deviation of samples of similar breeding population planted on different soils. Asparagine and dopamine in nature serve distinctive functions for plant species. Asparagine is a common form of nitrogen source in vascular bundles (Lea et al. 2007) while dopamine is a non-protein amino acid with more sophisticated functions such as stimulating stress responses (Vranova et al. 2011). Nevertheless, both metabolites are indispensable for plant survival. Since no single protocol can capture the extensive plant metabolome, the described method provides a good approach for reducing the complexity of the biological samples with LC-MS as one of the latest developed technologies for oil palm metabolomics. This attempt will hopefully pave the way for future studies to systematically link the activities in the field with

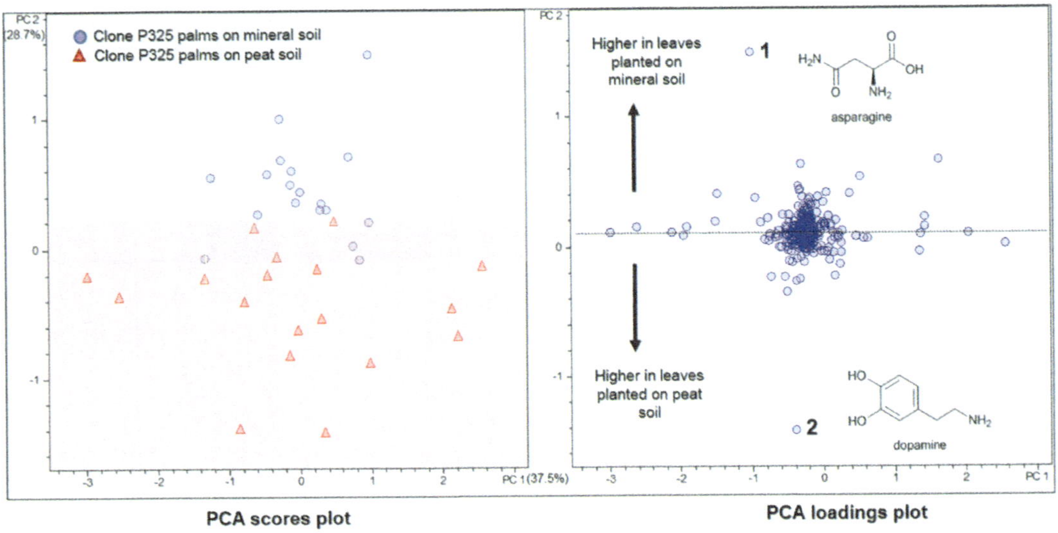

Fig. 10.3 Principal component analysis (PCA) scores and loading plots of oil palm leaf metabolome planted on different soil trials (adapted from Tahir et al. 2016)

chemometrics, facilitating discovery of information for oil palm agronomists and physiologists to further elevate oil palm productivity.

Biochemical studies on oil biosynthesis found oil deposition in the oil palm fruit mesocarp starting at approximately 15 weeks after anthesis (WAA) and continuing to fruit maturity at about 20 WAA (Oo et al. 1985). Comparative transcriptome and metabolome analyses on fatty acid and triacylglycerol biosynthetic pathways were able to map out metabolic differences associated with the mesocarp oil content and fatty acid composition (Wong et al. 2017; Teh et al. 2014; Dussert et al. 2013; Neoh et al. 2013; Bourgis et al. 2011; Tranbarger et al. 2011). Systems biology of oil palm fruit development can be studied through combined transcript, protein and metabolite analyses to uncover complex regulatory interactions. Proteome profiling takes us one step further towards molecular insights into fruit metabolism and oil accumulation (Ooi et al. 2015; Hassan et al. 2019; Loei et al. 2013). Proteome and metabolome assessment has been undertaken to understand the underlying developmental shifts during oil palm fruit development from 10 to 20 WAA (Hassan et al. 2016). Over-expressed proteins and differentially abundant metabolites indicate their significant roles and status during mesocarp tissue maturation. Proteins involved in energy metabolism (enolase and fructose-bisphosphate aldolase) and lipid metabolism were highly accumulated at the beginning of the fruit maturation process (from 15 WAA). Proteins abundant during the ripening phase (20 WAA) were involved in lipid metabolism (the fatty acid condensing enzymes), secondary metabolism (trans-resveratrol di-O-methyltransferase-like), cell wall biogenesis/degradation and plant defence (β-1,3-glucanase and glucan endo-1,3-beta-glucosidase-like) (Hassan et al. 2016).

Oil palm performance can be further improved by tailoring the fatty acid compositions to the requirements of the food and oleochemical industries. Oil palm fatty acid metabolism can be modified by crossing different oil palm types or species (conventional plant breeding), gene editing and metabolic engineering (Parveez et al. 2015; Ramli et al. 2009; Singh et al. 2008; Ramli et al. 2002a; Ramli et al. 2002b) to increase/reduce specific fatty acids. This would produce palm oil with even greater versatility and higher market value. Although the biochemistry of fatty acid biosynthesis in oil palm has been much studied (Dussert et al. 2013; Teh et al. 2013; Tranbarger et al. 2011; Sundram et al. 2003; Sambanthamurthi et al. 2000; Sambanthamurthi et al. 1999; Oo et al. 1985), crosstalk between transcriptional and metabolic controls in regulating fatty acid compositions is not well understood. In recent years, peptide-centric proteomics such as matrix-assisted laser ionization mass spectrometry (MALDI)-time-of-flight (TOF)/TOF, GeLC-MS/MS and two-dimensional (2D)-LC-MS/MS have been used to study proteins involved in the regulation (Lau et al. 2017; Hassan et al. 2019; Ooi et al. 2015; Lau et al. 2015; Loei et al. 2013). In addition to the upregulation of specific fatty acid regulatory genes such as WRINKLED1 (WRI1), another promising finding indicated that post-translational modification (PTM) might also be part of the regulation machinery (Lau et al. 2016). Several key proteins that are involved in the fatty acid biosynthesis mechanism have been shown to be differentially expressed during the peak of fatty acid production (Ooi et al. 2015; Loei et al. 2013). The differences in expression of these fundamental proteins were also observed between oil palm fruits with oils of different unsaturation (Lau et al. 2017; Lau et al. 2015). However, elucidating the complex signalling pathway scan is challenging and laborious. Therefore, a combination of several omics approaches including proteomics to detect the dynamic changes is crucial and continuously being improved on to help oil palm improvement.

MPOB was the first to publish a customized method for proteomic analysis on oil palm plastid-derived proteins (Lau et al. 2016). The report described the link between post-translational protein modifications (phosphorylation) and regulation of fatty acid biosynthetic enzymes possibly by activation/deactivation of their protein conformations (Lau et al. 2016). Although plant and animal fatty acid biosynthesis have been investigated extensively, the less abundant enzymes involved in critical steps in oil palm fatty acid biosynthesis have never been reported. Integration of this information via omics technologies could potentially enhance the efforts to increase desired high-value fatty acids such as oleic acid and improve the marketability of palm oil globally. Together with the existing breeding programmes, coalesced genomic works, proteomics and metabolomics could help to establish the ideal ecosystem for biomarker discovery. Information from systematic and cohesive omics experiments will piece together the systems biology of fatty acid production in attempts to boost higher oil yield as well as strengthen the nutritional advantages of palm oil globally.

10.4 Deciphering Basal Stem Rot (BSR) Disease by Proteomics and Metabolomics for Early Detection in Infected Palms

The oil palm industry in Malaysia is being threatened by basal stem rot (BSR) caused by the fungus *Ganoderma*. The disease has been reported to attack oil palm trees planted on all types of soil such as laterite, peat, coastal and inland soils. BSR incidence is dire on oil palm grown on coastal soil especially those previously planted with coconut and/or oil palm which were left to decompose after death (Idris et al. 2011; Idris 2004). BSR was initially reported to infect long-standing oil palm trees of over 25 years, but during the 1950s, the disease spread to younger palms of 10–15 years old (Turner and Gillbanks 2003). BSR is currently considered an epidemic oil palm disease in the country, sometimes occurring on young palms just 1–2 years after planting (Idris et al. 2011). No effective control is known although ameliorative treatments such as soil mounding, sanitation by removal of diseased palms tissues, stump treatment with dazomet, application of fungicide, e.g. hexaconazole, biological treatment and application of the recently developed GanoCareTM fertilizer (Idris et al. 2011) appear to offer some respite.

External symptoms of BSR in young palms comprise yellowing or mottling of the lower fronds which eventually become necrotic (Rees et al. 2009). Infection produces unfolded young leaves that become chlorotic and may be reduced in length, sometimes with necrotic tips. As the disease progresses, the most obvious symptoms include pale appearance with retarded growth and infected spear leaves remain unopened.

Similar symptoms are observed on mature palms, with multiple unopened spear leaves and a generally pale leaf canopy. Infected palms usually die within 6–12 months of the appearance of unexpanded spear leaves with necrosis in the oldest fronds extending to younger regions of the crown. At any time or point in the course of the disease symptoms, the appearance of basidioma firmly establishes the presence of infection and as the disease severity increases, the basal stem will rot and the palm will eventually fall or remain standing as a bare trunk/stump. Palms normally die within 1–3 years after onset of the foliar symptoms (Idris et al. 2004).

The limited understanding of *Ganoderma* infection has hampered the disease management and control (Azahar et al. 2014; Susanto et al. 2005). However, omics can provide a means to study the palm-pathogen interaction at the cellular and biochemical levels. In recent years, there have been an increased number of metabolomics and proteomics investigations to portray the different pathological states of plant disease. A reliable workflow is much needed to outline the BSR disease mechanism and to formulate diagnostic assays for early detection as well as to develop agronomical biomarkers to rate the oil palm tolerance to *Ganoderma*. With omics, greater understanding of the oil palm biochemical and cellular response to this disease can be gained along with the identification of biomarker(s) for early detection.

Since the route of *Ganoderma* infection is still unclear, extensive efforts were initiated to map the global transcriptome, protein and metabolome expression within the diseased palm. A global view of the oil palm leaf proteome of oil palm seedlings, both uninoculated and inoculated with *Ganoderma*was obtained by analysing protein spots that changed in abundance at two and three months post-inoculation using LC-MS (Daim et al. 2015). Based on classification using the Kyoto Encyclopedia of Genes and Genomes (KEGG) database, leaf proteins of lower abundance are involved in photosynthesis, immunity and defence response, energy production, responses to stress, energy and nitrogen metabolism while proteins found in higher abundance are involved in carbohydrate metabolism.

Crop-pathogen interaction has also been widely studied via gel-based approaches in order to identify proteins in relative abundance between physiological states of diseased and healthy crops, for example the leaf rust disease in wheat (Rampitsch et al. 2006), rice blast disease (Shenton et al. 2012) and Asian soybean rust disease (Ganiger et al. 2013). In oil palm, the proteomics approach has been employed to investigate the biochemical and cellular processes during BSR disease (Daim et al. 2015; Syahanim et al. 2013; Jameel et al. 2014). As mentioned earlier, the mode of infection of *Ganoderma* in oil palm is still unclear. The expression of defence-related genes can be observed in both root and leaf tissues of *Ganoderma*-infected seedlings (Tee et al. 2013; Alizadeh et al. 2011). In comparison with the symbiotic *Trichoderma harzianum*, inoculation with pathogenic *Ganoderma boninense* was found to induce increased expression of stearoyl-acyl carrier protein desaturase (SAD) 1 and type 3 metallothionein (MT3)-A in roots and SAD2 in leaves at 21 days after treatment prior to any appearance of any physical symptoms with the transcripts of MT3 genes being detected at a later time (day-42) in the leaf tissues (Alizadeh et al. 2011).

Ganoderma infection through the plant root system has been investigated using the rubber wood block method (Breton et al. 2006; Hasan and Turner 1998; Sariah et al. 1994; Lim et al. 1992; Navaratnam and Chee 1965). Transcriptome analysis based on cDNA microarray of root samples from uninoculated oil palm seedlings and those artificially inoculated with *Ganoderma* using the rubber wood block revealed 61 differentially regulated transcripts that correspond to genes encoding isoflavone reductase, early-methionine (Em) protein H2, SYG1/Pho81/XPR1 (SPX) domain-containing protein 1, pathogenesis-related protein 1 and vicilin-like antimicrobial peptide (Tee et al. 2013). Information from a transcriptome survey on oil palm root tissue treated with *G. boninense* enabled the postulation of the contagion response mechanisms to formulate better strategies for prevention and management of BSR (Ho et al. 2016), as summarized in Table 10.1.

The oil palm root system is the first line of defence against soil pathogens, including *Ganoderma*. Therefore, to improve our understanding on the mechanisms involved during plant-pathogen interaction, omics analysis on this important organ of the infected palm is valuable. Proteome analysis of root extracts of oil palm seedlings following artificial infection was performed to determine the alteration in abundance of proteins (Syahanim et al. 2013). 2D gel electrophoresis coupled with MALDI mass spectrometry analysis performed on the digested peptides, and the results matched with in-house database. Beta-1,3- glucanase, glutathione S-transferase, early flowering protein 1, nucleoside disphosphate kinase, thioredoxin H2 and ferritin were identified in the infected palm root at day-7 post-infection. These proteins have been previously reported in plant defence and stress responses. Changes in abundance of the oil palm root proteome at day-14 post-infection were reported in another investigation where a total of two proteins, namely malate dehydrogenase and cysteine synthase were in high abundance and five proteins, enolase, fructokinase 1 and 2, ATP synthase and catechol *O*-methyltransferase were reduced after *Ganoderma* infection. These proteins appear to be interconnected via metabolic and defence pathways (Al-Obaidi et al. 2014).

Comparison of the protein inventories of pathogenic and non-pathogenic *Ganoderma* species may provide useful information on the molecular mechanisms involved in fungal pathogenicity. A comprehensive analysis of proteins from *Ganoderma* species will allow for the evaluation of the phenotypic differences between them. Indirectly, characterization of the *Ganoderma* species can also assist in achieving effective disease control management for BSR. A wide protein profiling from the mycelia of *Ganoderma* spp has been initiated (Dzulkafli et al. 2016). Two approaches were carried out in this experiment concurrently for label-based and label-free proteomics. Protein extraction protocols for sodium dodecyl sulphate-polyacrylamide gel electrophoresis (SDS-PAGE) analysis of *Ganoderma* spp. using suitable extraction buffer resulted in high protein yield have also been developed with successful separation of proteins according to their molecular weights. An effective strategy is being formulated to rapidly identify the proteins from *Ganoderma* spp. to gain extensive information on the fungus pathogenicity that will allow for strategies to be formulated for better control.

Table 10.1 Oil palm response mechanisms towards *G. boninense*

No.	Infection corollary of host towards pathogen	Possible manifestation
1	Oil palm root defence responses towards fungal cell wall components	Presence of ergosterol in oil palm tissue
2	Suppression of jasmonic acid (JA) biosynthesis, signalling and JA-mediated defence	Decreased concentration of JA
3	Activation of salicylic acid (SA) biosynthesis, signalling and SA-mediated defence	Increased concentration of SA
4	Competition between pathogen and host in transcriptional regulation of ethylene biosynthesis, hydrogen peroxide production and scavenging, apoptosis and autophagy	Physical disease symptoms
5	Antifungal activities against *G. boninense* in infected root tissue	Production of pathogenesis-related proteins (chitinases, glucanases, defensins, protease inhibitors) and secondary metabolites
6	Reinforcement of host cell wall through cell wall modifying enzymes	Increased expression of cell wall modifying enzymes

10.4.1 Omics Towards Predicting Tolerant Palms

Over the years, several measures have been used to control BSR in oil palm (Idris et al. 2011). Some are treatments applied to the visibly infected palms to slow down the disease progression including biological agents and systematic fungicides (Susanto et al. 2005; Idris et al. 2010). Other measures are the establishment of biomarkers to fully characterize the palm cellular and metabolic mechanisms, which, in the long run, would allow for the breeding and selection of palms tolerant to BSR. Screening of planting materials from various sources with a very wide genetic base has revealed the susceptibility of Deli oil palm, and good performance of African materials in nursery screening in Malaysia and in the field in Indonesia (Durand-Gasselin et al. 2005; Idris et al. 2004).

The ability of metabolomics to identify phenotypes of disease resistance traits is also expected to have a strong impact on biomarker discovery and validation of oil palm systems biology research. Identification of oil palm materials tolerant to *Ganoderma* is crucial as they portend a solution to the disease. Metabolomics combined with other strategies in the omics cascade may solve the problems in breeding for BSR resistance. Progress on plant metabolomics research relies largely on the choice of technologies used to comprehensively identify, quantify and localize every metabolite within (Hong et al. 2016). A series of integrated technologies and methodologies based on mass spectrometry (MS) including gas chromatography-mass spectrometry (GC-MS) have been applied to characterize potential tolerant palms for BSR (Rozali et al. 2017). The first application of LC-MS reported by Nurazah et al. (2017) on metabolite profiling and coupled with multivariate statistical analysis have revealed three metabolites from different classes of plant compounds—sugars, phenolic acids and organic acids—as potential oil palm metabolite markers that may assist breeders for the selection of partial tolerant oil palms (Table 10.2).

The ongoing efforts in explaining and elucidating the metabolic responses from primary and secondary metabolism in oil palm of diverse genetic backgrounds against BSR imply that metabolomics-assisted breeding can be useful in identifying palms more tolerant to the disease. The important role of metabolomics in breeding for disease resistance will become increasingly clear in the future. Methods have been developed for screening of highly tolerant palms to *Ganoderma boninense* via proteomics and metabolomics (Ramli et al. 2016). The proteins or metabolites identified from the highly tolerant genotype will not only serve as a base in the quest of novel defence compounds, but also as markers for characterization of the palm defensive state. The latter is especially useful in agronomic applications where useful trait specific markers are essential for crop protection. For example, identification of proteins and metabolites from the infected palms during early stages of BSR infection will provide a better understanding of resistance gene functions and the ability to formulate early-detection diagnostic kits for the disease (Mohamad Arif et al. 2007; Syahanim et al. 2013).

Comparative omics analysis between oil palm inoculated with pathogenic *Ganoderma* sp. and those inoculated with biocontrol fungi, such as *Trichoderma* sp. and non-pathogenic *Ganoderma*, (*G. tornatum*) may unravel the response

Table 10.2 Fold changes in concentrations of the three putative metabolites

m/z (mass to charge)	Putative metabolites	Fold changes* (Deli *dura*/MPOB-Cameroon *dura*)
173.0453	Shikimic acid	6.8781
179.0558	Glucose	1.6034
133.0143	Malic acid	1.9811

Note *Significant at $p < 0.01$. Adapted from Nurazah et al. (2017)

factors and possibly enable identification of *Ganoderma* tolerant oil palm planting materials. A recent transcriptome profiling of oil palm roots treated with *T. harzianum* identified upregulated genes related to transportation including transport of amines, nitrogenous compounds and carboxylic acid. These were not found in oil palm roots treated with *Ganoderma*. The gene encoding polygalacturonase was down-regulated in the *T. harzianum*-treated oil palm roots, indicating distinct cell wall alteration mechanisms in the root tissues stimulated by the fungi (Ho et al. 2016). Continuing this investigation on other omics platforms should uncover more facets of the *Ganoderma*-oil palm interaction possibly allowing for eventual formulation of reliable screening methods for tolerance to the disease. Then, further omics work may finally differentiate the response mechanisms of tolerant and susceptible oil palm to pathogenic *Ganoderma*.

10.5 Integrative Omics and Informatics for Oil Palm

The complexities and diverse physiological properties of oil palm such as yield, disease resistance and adaptation to abiotic stresses require integrated omics strategies for exploring the metabolic and regulatory mechanisms to differentiate the phenotypes. For instance, a sensitive and efficient proteomic platform will facilitate profiling the proteins for photosynthesis, metabolism, cellular biogenesis, stress response and other biological processes transport using mass spectrometry. The data analysis is then facilitated by using National Center for Biotechnology Information (NCBI)-proteins and Malaysian Palm Oil Board (MPOB) oil palm expressed sequence tag (EST) databases (Tan et al. 2017). Another study on oil palm seedlings challenged with water dearth reported total soluble protein, chlorophyll content, proline, antioxidant enzyme activities and expression of stress-related genes as possible drought-response biochemical-molecular indicators (Azzeme et al. 2016). Metabolome profiling of similar tissues uncovered a score of several flavone glycosides, sugars and fatty acids which may have nutraceutical applications due to their bioactive properties (Tahir et al. 2012). In reflection of these examples, the acquisition of in-depth genome, proteome and metabolome information using high throughput technology such as mass spectrometry combined with methodical assessment of biological systems under different conditions such as stress would provide better if not ideal description of the oil palm systems biology. Despite the fact that biologists are familiar with the central dogma of molecular biology, the translation of genomic information into protein abundance is not as straightforward as in theory (Maier et al. 2009). Furthermore, even in a systematic comparison, the correlation between mRNA expression levels and protein levels is scarce and insufficient for genome-wide protein dynamics and biochemical regulation cognizance without metabolite profiling (Wienkoop et al. 2008). Thus, analysis of an individual omics would probably be a conjecture if used for describing interrelation and the functioning of larger systems.

The advancement of each research modus of genomics, transcriptomics, proteomics, metabolomics and phenomics has been at its own pace and influenced by the availability and improvement of analytical instruments, reagents and also data analysis techniques. Genomics for instance, was revolutionized by next-generation sequencing (NGS) technologies that have reduced time and cost of research (Metzker 2010). On the other hand, exploration of the proteome and metabolome is gaining momentum aided by several informatics platforms such as databases, function annotation and pattern recognition statistical tools. Inherently, the high-throughput technologies used in each omics platform generates large volumes of data that call for efficient informatics solutions. Multi-omics data can be analysed and interpreted either independently or in an integrated manner, with both options bearing considerable challenges of their own, e.g. large data volumes and data-type diversity and dimensionality. Using metabolomics as a base due to its multifaceted interaction between molecular omics, environment and the phenotype, binary paired omics analysis of genomics, transcriptomics and proteomics versus metabolomics data can be performed using canonical correlation and network analysis (Krumsiek et al. 2016). For integrative analysis of multi-omic layers, formulation of an effective method of data analysis involves at least two criteria: the use of networks and Bayesian theory (Chasman et al. 2016; Bersanelli et al. 2016; Sass et al. 2013). In the long run, information recorded from measurements and observation of a biological system can be used to predict its phenotypic traits and the consequences of interest such as disease status and responses to a stimulus or treatment and to facilitate molecular signature discovery (Acharjee et al. 2016; Sung et al. 2012) by converting explanatory models to predictive models. The algorithms for the construction of frameworks and models for integrative multi-omic informatics are usually implemented using R programming language available at http://www.R-project.org (Bersanelli et al. 2016; Fondi and Liò 2015).

For a comprehensive systemic view, omics data can then be gauged by placing them in networks or pathways for visualization purposes and phenomics acumens. One example of an all-embracing portal for omics analyses and databases is the European Bioinformatics Institute (EMBL-EBI) online services at http://www.ebi.ac.uk/ that provide comprehensive and user-friendly tools and archives for the public (Cook et al. 2016). The PRoteomics IDEntifications database (http://www.ebi.ac.uk/pride) of protein and peptide identifications is a comprehensive collection of proteomics identification and metadata categorized according to species, tissue, experiments (including mass spectrometry imaging) and several other browsing parameters (Römpp et al. 2015; Jones et al. 2006; Martens et al. 2005). In similar fashion, MetaboLights was developed as a cross-species, cross-application metabolome data repository of EBI which incorporate details on raw experimental information and relevant metadata (Haug et al. 2013). Following the dissemination publication of the maize (*Zea mays*) genome sequence in 2009, post-genomics investigations concerning its yield and biomass gained extra momentum to tackle challenges relevant to the oil palm industry such as increased food and (bio)fuel demand and climate change. The transcriptomics, metabolomics, ionomics, proteomics and phenomics data of this important food and fodder species can be accessed freely at the OPTIMAS (OPTImization of bioMASs) Data Warehouse (OPTIMAS-DW) (http://www.optimas-bioenergy.org/optimas_dw), an omics databank developed by Friedrich–Alexander University of Erlangen–Nuremberg, Max Planck Institute of Molecular Plant Physiology and Leibniz Institute of Plant Genetics and Crop Plant Research (Colmsee et al. 2012). Data for the different omics domains were obtained from the same plant material to allow direct correlation and functional interpretation of maize systems biology.

Large amounts of metabolome data are generated today and with new generation technologies becoming more common in laboratories; handling and managing of the datasets are a big challenge. With the complexity and the intricacy of generated metabolome data, it is important to develop systems which allow for large data deposition in a systematic manner. A database for oil palm metabolome with functions that facilitate query and retrieval of any desired data for dissemination purposes was constructed to circumvent challenges of conventional data keeping (Nur-Ain et al. 2015). It is structured based on the research project and comprises study descriptions, researcher details, material resources, published references, methodology and utilized analytical techniques such as mass spectrometry (MS), ultra violet (UV) detection and nuclear magnetic resonance (NMR) spectroscopy. It also contains metabolite descriptions including their IUPAC name, common name, molecular formula, molecular weight, molecular structure, mass-to-charge ratios (m/z ratio) and other relevant information. Construction of this library is expedient due to its capacity to store in different formats such as text, numbers and images (molecular structure and MS fragments spectra).

At the moment, the metabolome collection includes in-house entries of several research investigations contributing to different applications such as nutraceuticals and plant pathology (Rozali et al. 2017; Nurazah et al. 2013; Tahir et al. 2012). Users are given several options of library searching utility which include text search and molecular structure search by uploading the molecular structure file, drawing via PubChem Sketcher and simplified molecular-input line-entry system (SMILES) string. In addition to this, there are browsing strategies for those who prefer a quick check for relevant information in terms of species, oil palm plant parts and instrument platform. The library can be viewed using the current Internet browsers.

The metabolome database offers an informative public resource to the scientific community and oil palm industry players for their use whether in systems biology or commercial endeavours. This database is the first piece of the oil palm omics informatics ensemble, and the streamlining of the genomic and proteomic information into one complete platform will follow suit.

10.6 Concluding Remarks

Application of the various omics platforms to analyse complex processes at different levels of oil palm tissue metabolic events can provide new insights into oil palm biology which will be useful for the crop improvement and the development of new and improved oil palm varieties. The progress of 'omics' technologies has opened up avenues to generate a vast amount of datasets for this crop species. Integration of genome and functional omics data such as transcriptomics, proteomics and metabolomics with phenotypic information is a remarkable development leading towards understanding of factors responsible for important agronomic traits. The major focus in oil palm research are factors for high oil yield, modifying the fatty acid biosynthesis process for tailored oil composition and confronting the *Ganoderma* disease. The oil palm industry can now access the so-called modern omics tools to obtain meaningful understanding of the biological information towards the improvement of this outstanding crop. The best planting materials from the breeders matched with good agronomical practices by understanding the response of the oil palm to its biotic and abiotic stimuli will confer a productivity leap for the industry in overcoming its current and upcoming challenges to remain as the country's socio-economic driver.

Acknowledgements The authors wish to thank the Director-General of MPOB for permission to publish this chapter.

References

Acharjee A, Kloosterman B, Visser RGF, Maliepaard C (2016). Integration of multi-omics data for prediction of phenotypic traits using random forest. BMC Bioinform 17(Suppl 5) 180:363–373

Alexandersson E, Jacobson D, Vivier MA, Weckwerth W, Andreasson E (2014) Field-omics-understanding large-scale molecular data from field crops. Front Plant Sci 5(286):1–6

Alizadeh F, Abdullah SNA, Khodavandi A, Abdullah F, Yusuf UK, Chong PP (2011) Differential expression of oil palm pathology genes during interactions with *Ganoderma boninense* and *Trichoderma harzianum*. J Plant Physiol 168:1106–1113

Alm EKH (2007) Success stories of agricultural long-term experiments. In: Kungl. Skogs- Och Lantbruksakademiens Tidskrift, 146. ISBN: 978-91-85205-61-5

Al-Obaidi J, Mohd-Yusuf Y, Razali N, Jayapalan J, Tey C-C, Md-Noh N, Junit S, Othman R, Hashim O (2014) Identification of proteins of altered abundance in oil palm infected with *Ganoderma boninense*. Int J Mol Sci 15:5175–5192

Azahar TM, Idris AS, Abu Hassan D, Norazlin I (2014) Assessment of basal stem rot disease distribution in palm oil plantation using geographical information system. J Sci Technol 81–92

Azzeme AM, Abdullah SNA, Aziz MA, Wahab PEM (2016) Oil palm leaves and roots differ in physiological response, antioxidant enzyme activities and expression of stress-responsive genes upon exposure to drought stress. Acta Physiol Plant 38(52):1–12

Barcelos E, Rios SD, Cunha RN, Lopes R, Motoike SY, Babiychuk E, Skirycz A, Kushnir S (2015) Oil palm natural diversity and the potential for yield improvement. Front Plant Sci 6(190):1–16

Bersanelli M, Mosca E, Remondini D, Giampieri E, Sala C, Castellani G, Milanesi L (2016) Methods for the integration of multi-omics data: mathematical aspects. BMC Bioinform 15:167–177

Billotte N,Marseillac N,Risterucci AM, Adon B, Brottier P, Baurens FC, Singh R, Herran A, Asmady H, Billot C, Amblard P, Durand-Gasselin T, Courtois B, Asmono D, Cheah SC, Rohde W, Ritter E, Charrier A

(2005) Microsatellite-based high density linkage map in oil palm (*Elaeis guineensis* Jacq.). Theor Appl Genet 110:754–765

Bird A (2002) DNA methylation patternsand epigenetic memory. Genes Dev 16(1):6–21

Bourgis F, Kilaru A, Cao X, Ngando-Ebongue G, Drira N, Ohlrogge JB (2011) Comparative transcriptome and metabolite analysis of oil palm and date palm mesocarp that differ dramatically in carbon partitioning. Proc Natl Acad Sci 108(30):12527–12532

Breton F, Hasan Y, Hariadi S, Lubis Z, De Franqueville H (2006) Characterization of parameters for the development of an early screening test for basal stem rot tolerance in oil palm progenies. J Oil Palm Res (Special Issue):24–36

Chasman D, Siahpirani AF, Roy S (2016) Network-based approaches for analysis of complex biological systems. Curr Opin Biotech 39:157–166

Cheah SC, Abdullah SNA, Ooi LCL, Rahimah AR, Maria M (1993) Detection of DNA variability in the oil palm using RFLP probes. 1991 PORIM international palm oil conference—agriculture (module 1). Kuala Lumpur, Malaysia, pp 144–150

Chen R, Snyder M (2013) Promise of personalized omics to precision medicine. Wiley Interdiscip Rev Syst Biol Med 5(1):73–82

Chong KL, Kanniah KD, Pohl C, Tan KP (2017) A review of remote sensing applications for oil palm studies. Geo Spatial Inform Sci 20(2):184–200

Colmsee C, Mascher M, Czauderna T, Hartmann A, Schlüter U, Zellerhoff N, Schmitz J, Bräutigam A, Pick TR, Alter P, Gahrtz M, Witt S, Fernie AR, Börnke F, Fahnenstich H, Bucher M, Dresselhaus T, Weber APM, Schreiber F, Scholz U, Sonnewald U (2012) OPTIMAS-DW: a comprehensive transcriptomics, metabolomics, ionomics, proteomics and phenomics data resource for maize. BMC Plant Biol 12 (245):1–10

Cook CE, Bergman MT, Finn RD, Cochrane G, Birney E, Apweiler R (2016) The European Bioinformatics Institute in 2016: data growth and integration. Nucleic Acids Res 44(Database):D20–D26

Corley RHV (2009) How much palm oil do we need? Environ Sci Policy 12:134–139

Daim LDJ, Tek Ooi, Ithnin N, Mohd Yusof H, Kulaveerasingam H, Majid NA, Karsani SA (2015) Comparative proteomic analysis of oil palm leaves infected with *Ganoderma boninense* revealed changes in proteins involved in photosynthesis, carbohydrate metabolism, immunity and defense. Electrophoresis 36(15):1699–1710

Durand-Gasselin T, Asmady H, Flori A, Jacquemard JC, Hayun Z, And Breton F, De Franqueville H (2005) Possible sources of genetic resistance in oil palm (*Elaeis guineensis* Jacq.) to basal stem rot caused by *G. boninense*-prospects for future breeding. Mycopathologia 159:93–100

Dussert S, Guerin C, Andersson M, Joët T, Tranbarger TJ, Pizot M, Sarah G, Omore A, Durand-Gasselin T, Morcillo F (2013) Comparative transcriptome analysis of three oil palm fruit and seed tissues that differ in oil content and fatty acid composition. Plant Physiol 162:1337–1358

Dzulkafli SB, Abrizah O, Benjamin LYC, Syahanim S, Idris AS, Ramli US (2016) Optimization of protein extraction from *Ganoderma boninense* for SDS PAGE analysis. Trans Persatuan Genetik Malaysia 3:193–197

Fiehn O (2002) Metabolomics—the link between genotypes and phenotypes. Plant Mol Biol 48:155–171

Fiocchi C (2014) Integrating omics: the future of IBD? Dig Dis 32(Suppl 1):96–102

Fondi M, Liò P (2015) Multi-omics and metabolic modelling pipelines: challenges and tools for systems microbiology. Microbiol Res 171:52–64

Ganiger M, Walker DR, Chen ZY (2013) Proteomics based study of soybean and *Phakopsora pachyrhizi* interaction. In: Proceedings of the Twelfth I. E. Melhus Graduate Student Symposium. Annual Meeting of the American Phytopathological Society (APS), 6 August 2012 in Providence, RI. Plant Health Prog, pp 1–13. https://doi.org/10.1094/php-2013-1125-01-rs

Goodacre R, Vaidyanathan S, Dunn WB, Harrigan GG, Kell DB (2004) Metabolomics by numbers: acquiring and understanding global metabolite data. Trends Biotech 22(5):245–252

Han Y, Garcia BA (2013) Combining genomic and proteomic approaches for epigenetics research. Epigenomics 5(4):439–452

Hasan Y, Turner PD (1998) The comparative importance of different oil palm tissue as infection sources of basal stem rot in replantings. Planter 74(864):119–135

Hassan H, Tahir NI, Ramli US (2016) Proteome and metabolome assessment of oil palm fruit development for advanced breeding perspective. Trans Persatuan Genetik Malaysia 3:199–204

Hassan H, Mohd Din A, Weckwerth W and Ramli US (2019) Deciphering key proteins of oil palm (*Elaeis guineensis* Jacq.) fruit mesocarp development by proteomics and chemometrics. Electrophoresis 40 (2):254–265

Haug K, Salek RM, Conesa P, Hastings J, De Matos P, Rijnbeek M, Mahendraker T, Williams M, Neumann S, Rocca-Serra P, Maguire E, González-Beltrán A, Sansone S, Griffin JL, Steinbeck C (2013) MetaboLights -an open-access general-purpose repository for metabolomics studies and associated meta-data. Nucl Acids Res 41(D1):D781–D786

Ho CL, Tan YC, Yeoh KA, Ghazali AK, Yee WY, Hoh CC (2016) *De novo* transcriptome analyses of host-fungal interactions in oil palm (*Elaeis guineensis* Jacq.). BMC Genom 17(66):1–19

Hong J, Yang L, Zhang D, Shi J (2016) Plant metabolomics: an indispensable system biology tool for plant science. Int J Mol Sci 17(767):1–16

Idris AS, Kushairi A, Ismail S, Ariffin D (2004) Selection forpartial resistance in oil palm progenies to *Ganoderma* basal stem rot. J Oil Palm Res 16(1):12–18

Idris AS, Nasyarudin MNM, Maizatul SM, Zaiton S (2010) Gano EB1—A bacterial biocontrol agent for *Ganoderma* in oil palm. MPOB Inf. Ser. MPOB TT No. 443

Idris AS, Mior MHAZ, Maizatul SM, Kushairi A (2011) Survey on status of *Ganoderma* disease of oil palm. In: Proceeding of MPOB International Palm Oil Conference (PIPOC 2011), Malaysian Palm Oil Board, Kuala Lumpur, Malaysia, 15–17 November 2011, pp 235–238

Jaligot E, Beulé T, Rival A (2002) Methylation-sensitive RFLPs: characterisation of two oil palm markers showing somaclonal variation-sensitive associated polymorphism. Theor Appl Genet 104:1263–1269

Jones P, Côté RG, Martens L, Quinn AF, Taylor CF, Derache W, Hermjakob H, Apweiler R (2006) PRIDE: a public repository of protein and peptide identifications for the proteomics community. Nucleic Acids Res. 34(Database issue):D659–D663

Keurentjes JJ (2009) Genetical metabolomics: closing in on phenotypes. Curr Opin Plant Biol 2:223–230

Krumsiek J, Bartel J, Theis FJ (2016) Computational approaches for systems metabolomics. Curr Opin Biotechnol 39:198–206

Lau BYC, Clerens S, Morton JD, Dyer JM, Deb-Choudhury S, Ramli US (2015) Method developments to extract proteins from oil palm chromoplast for proteomic analysis. Springer Plus 4:791

Lau BYC, Clerens S, Morton JD, Dyer JM, Deb-Choudhury S, Ramli US (2016) Application of mass spectrometry approach to detect the presence of fatty acid biosynthetic phosphopeptides. Protein J 35:163–170

Lau BYC, Morton JD, Deb-Choudhury S, Clerens S, Dyer JM, Ramli US (2017) Differential expression analysis of oil palm fatty acid biosynthetic enzymes with gel-free quantitative proteomics. J Oil Palm Res 29(1):23–34

Lea PJ, Sodek L, Parry MAJ, Shewry R, Halford NG (2007) Asparagine in plants. Ann Appl Biol 150:1–26

Lim T, Chung G, Ko W (1992) Basal stem rot of oil palm caused by *Ganoderma boninense*. Plant Pathol Bull 1 (3):147–152

Loei H, Lim J, Tan M, Lim TK, Lin QS, Chew FT, Kulaveerasingam H, Cung MC (2013) Proteomic analysis of the oil palm fruit mesocarp reveals elevated oxidative phosphorylation activity is critical for increased storage oil Production. J Proteome Res 12:5096–5109

Maier T, Güell M, Serrano L (2009) Correlation of mRNA and protein in complex biological samples. FEBS Lett 583:3966–3973

Malaysian Palm Oil Board (MPOB) (2011) Palm oil: the way forward. Entry

Malaysian Standard MS 2099 (2008) Oil palm clones for commercial planting—specification for ortet selection. Department of Standards Malaysia

Mann M, Kelleher NL (2008) Precision proteomics: the case for high resolution and high mass accuracy. PNAS 105(47):18131–18138

Martens L, Hermjakob H, Jones P, Adamski M, Taylor C, States D, Gevaert K, Vandekerckhove J, Apweiler R (2005) PRIDE: the proteomics identifications database. Proteomics 5:3537–3545

Mayes S, James C, Horner SF, Jack PL, Corley RHV (1996) The application of restriction fragment length polymorphism for the genetic fingerprinting of oil palm (*Elaeis guineensis*). Mol Breed 2:175–180

Mayes S, Jack PL, Marshall D, Corley RHV (1997) Construction of a RFLP genetic linkage map for oil palm (*Elaeis guineensis* Jacq.). Genome 40:116–122

Mayes S, Jack PL, Corley RHV (2000) The use of molecular markers to investigate the genetic structure of an oil palm breeding programme. Heredity 85:288–293

Metzker ML (2010) Sequencing technologies—the next generation. Nat Rev Genet 11:31–46

Mohamad Arif AM, Abrizah O, Zetty Norhana BY, Syahanim S, Idris AS, Mohd Din A, Sambanthamurthi R (2007) Molecular and biochemical approaches to understanding oil palm-*Ganoderma* interactions. In: Proceeding of MPOB International Palm Oil Conference (PIPOC 2007), Kuala Lumpur Convention Centre, Kuala Lumpur, Malaysia, pp 228–246

Murphy DJ (2014) The future of oil palm as a major global crop: opportunities and challenges. J Oil Palm Res 26(1):1–24

Navaratnam SJ, Chee KL (1965) Root inoculation of oil palm seedlings with *Ganoderma* sp. Plant Dis 489:1011–1012

Neoh BK, Teh HF, Ng TLM, Tiong SH, Thang YM, Ersad MA, Mohamed M, Chew FT, Kulaveerasingam H, Appleton DR (2013) Profiling of metabolites in oil palm mesocarp at different stages of oil biosynthesis. J Agric Food Chem 61:1920–1927

Nur-Ain I, Nurazah Z, Rozali NL, Halim MA, Rosli R, Abrizah O, Ramli US, Tahir NI (2015) Construction of integrated oil palm metabolome chemical resources library. In: Proceeding of MPOB International Palm Oil Conference (PIPOC 2015), Kuala Lumpur Convention Centre, Kuala Lumpur, Malaysia, 6–8 October 2015, pp 381–385

Nurazah Z, Idris AS, Kushairi A, Ramli US (2013) Metabolite profiling of oil palm towards understanding basal stem rot (BSR) disease. J Oil Palm Res 25 (1):58–71

Nurazah Z, Idris AS, Kushairi A, Amiruddin MD, Abrizah O, Ramli US (2017) Metabolomics unravel differences between Cameroon dura and Deli dura oil palm (*Elaeis guineensis* Jacq.) genetic backgrounds against basal stem rot. J Oil Palm Res 29(2):227–241

OECD/FAO (2017) OECD-FAO Agricultural Outlook 2017–2026. OECD Publishing, Paris. http://dx.doi.org/10.1787/agr_outlook-2017-en

Oil World Annual (2017) Palm oil: world production (1000T), Yields (T/ha) and Mature Area (1000 ha). Oil World, ISTA Mielke GmbH, Hamburg, Germany

Ong-Abdullah M, Ordway JM, Jiang N, Ooi SE, Sarpan N, Azimi N, Hashim AT, Ishak Z, Rosli SK,

Malike FA, Bakar NA, Marjuni M, Abdullah N, Yaakub Z, Amiruddin MD, Nookiah R, Singh R, Low ET, Chan KL, Azizi N, Smith SW, Bacher B, Budiman MA, Van Brunt A, Wischmeyer C, Beil M, Hogan M, Lakey N, Lim CC, Arulandoo X, Wong CK, Choo CN, Wong WC, Kwan YY, Alwee SS, Sambanthamurthi R, Martienssen RA (2015) Loss of Karma transposon methylation underlies the mantled somaclonal variant of oil palm. Nature 525(7570):533–537

Oo KC, Teh SK, Khor HT, Ong ASH (1985) Fatty acid synthesis in the oil palm (*Elaeis guineensis*): incorporation of acetate by tissue slices of the developing fruit. Lipids 20(4):205–210

Ooi THE, Yeap WC, Daim LDJ, Ng BZ, Lee FC, Othman AM, Appleton DR, Chew FT, Kulaveerasingam H (2015) Differential abundance analysis of mesocarp protein from high- and low-yielding oil palms associates non-oil biosynthetic enzymes to lipid biosynthesis. Proteome Sci 13:28

Parveez GKA, Rasid OA, Masani MYA, Sambanthamurthi R (2015) Biotechnology of oil palm: strategies towards manipulation of lipid content and composition. Plant Cell Rep 34(4):533–543

Rajanaidu N, Kushairi A, Rafii M, Din M, Maizura I, Jalani B (2000) Oil palm breeding and genetic resources. In: Basiron YB, Jalani B, Chan KW (eds) Advances in oil palm research. Malaysian Palm Oil Board, Kuala Lumpur, pp 171–227

Ramli US, Baker DS, Quant PA, Harwood JL (2002a) Control mechanisms operating for lipid biosynthesis differ in oil-palm (*Elaeis guineensis* Jacq.) and olive (*Olea europaea* L.) callus cultures. Biochem J 364 (1):385–391

Ramli US, Baker DS, Quant PA, Harwood JL (2002b) Use of control analysis to study the regulation of plant lipid biosynthesis. Biochem Soc Trans 30:1043–1046

Ramli US, Salas JJ, Quant PA, Harwood JL (2009) Use of metabolic control analysis to give quantitative information on control of lipid biosynthesis in the important oil crop, *Elaeis guineensis* (oil palm). New Phytol 184(2):330–339

Ramli US, Lau BYC, Tahir NI, Shahwan S, Hassan H, Nurazah Z, Rozali NL, Dzulkafli S, Nur-Ain I, Abrizah O (2016) Proteomics and metabolomics: spearheading oil palm improvement and sustainability. Planter 92(1087):727–737

Rampitsch C, Bykova NV, Mccallum B, Beimcik E, Ens W (2006) Analysis of the wheat and *Puccinia triticina* (leaf rust) proteomes during a compatible host-pathogen interaction. Proteomics 6:1897–1907

Rees RW, Flood J, Hasan Y, Potter U, Cooper RM (2009) Basal stem rot of oil palm (*Elaeis guineensis*); mode of root infection and lower stem invasion by *Ganoderma boninense*. Plant Pathol 58(5):982–989

Römpp A, Wang R, Albar JP, Urbani A, Hermjakob H, Spengler B, Vizcaíno JA (2015) A public repository for mass spectrometry imaging data. Anal Bioanal Chem 407:2027–2033

Rozali NL, Yarmo MA, Idris AS, Kushairi A, Ramli US (2017) Metabolomics differentiation of oil palm (*Elaeis guineensis* Jacq.) spear leaf with contrasting susceptibility to *Ganoderma boninense*. Plant Omics 10:45–52

Sambanthamurthi R, Abrizah O, Ramli US (1999) Biochemical factors that control oil composition in the oil palm. J Oil Palm Res (Special Issue) 24–33

SambanthamurthI R, Sundram K, Tan YA (2000) Chemistry and biochemistry of palm oil. Prog Lipid Res 39:507–558

Sariah M, Hussin MZ, Miller RNG, Holderness M (1994) Pathogenicity of *Ganoderma boninense* tested by inoculation of oil palm seedlings. Plant Pathol 43 (3):507–510

Sass S, Buettner F, Mueller NS, Theis FJ (2013) A modular framework for gene set analysis integrating multilevel omics data. Nucleic Acids Res 41 (21):9622–9633

Sawada Y, Hirai MY (2013) Integrated LC-MS/MS system for plant metabolomics. Comput Struct Biotech J 4(5):1–6

Shenton MR, Berberich T, Kamo M, Yamashita T, Taira H, Terauchi R (2012) Use of intercellular washing fluid to investigate the secreted proteome of the rice—*Magnaporthe* interaction. J Plant Res 125:311–316

Singh R, Cheah SC (1999) Analysis of the inheritance of AFLP markers in an interspecific cross of oil palm using the pseudo-testcross strategy. J Oil Palm Res (Special Issue):64–73

Singh R, Zaki NM, Ting NC, Tan SG, Low ETL, Ithnin M, Cheah SC (2008) Exploiting an oil palm EST database for the development of gene-derived SSR markers and their exploitation for assessment of genetic diversity. Biologia 63(2):227–235

Singh R, Ong-Abdullah M, Low ETL, Manaf MAA, Rosli R, Nookiah R, Ooi LCL, Ooi SE, Chan KL, Halim MA, Azizi N, Jayanthi N, Bacher B, Lakey N, Smith SW, He D, Hogan M, Budiman MA, Lee EK, Desalle R, Kudrna D, Goicoechea JL, Wing RA Wilson RK, Fulton RS, Ordway JM, Martienssen RA, Sambanthamurthi R (2013a) Oil palm genome sequences reveals divergence of interfertile species in old and new worlds. Nature 500(7462):335–339

Singh R, Low ETL, Ooi LCL, Ong-Abdullah M, Ting NC, Jayanthi N, Nookiah R, Amiruddin Md, Rosli R, Manaf MAA, Chan KL, Halim MA, Azizi N, Lakey N, Smith SW, Budiman MA, Hogan M, Bacher B, Brunt AV, Wang C, Ordway JM, Sambanthamurthi R, Martienssen RA (2013b) The oil palm SHELL gene controls oil yield and encodes a homologue of SEEDSTICK. Nature 500(7462):340–344

Soh AC, Mayes S, Roberts JA (2017) Oil palm breeding. Genetics and genomics. CRC Press, USA

Sundram K, Sambanthamurthi R, Tan YA (2003) Palm fruit chemistry and nutrition. Asia Pacific J Clin Nutr 12(3):355–362

Sung J, Wang Y, Chandrasekaran S, Witten DM, Price ND (2012) Molecular signatures from omics data: from chaos to consensus. Biotech J 7:946–957

Susanto A, Sudharto PS, Purba RY (2005) Enhancing biological control of basal stem rot disease (*Ganoderma boninense*) in oil palm plantations. Mycopathologia 59(1):153–7

Syahanim S, Abrizah O, Manaf MAA, Idris AA, Amiruddin MD (2013) Identification of differentially expressed proteins in oil palm seedlings artificially infected with Ganoderma: a proteomics approach. J Oil Palm Res 25(3):298–304

Tahir NI, Shaari K, Abas F, Parveez GKA, Zamzuri I, Ramli US (2012) Characterization of apigenin and luteolin derivatives from oil palm (*Elaeis guineensis* Jacq.) leaf using LC-ESI-MS/MS. J Agric Food Chem 60(45):11201–11210

Tahir NI, Shaari K, Abas F, Zamzuri I, Tarmizi AH, Amiruddin MD, Parveez GKA, Ramli US (2016) Metabolome analysis of oil palm clone P325 of different planting trials. J Oil Palm Res 28(4):431–441

Tan HS, Jacoby RP, Ong-Abdullah M, Taylor NL, Liddell S, Wong WC, Chiew FC (2017) Proteomic profiling of mature leaves from oil palm (*Elaeis guineensis* Jacq.). Electrophoresis 38:1147–1153

Tee SS, Tan YC, Abdullah F, Ong-Abdullah M, Ho CL (2013) Transcriptome of oil palm (*Elaeis guineensis* Jacq.) roots treated with *Ganoderma boninense*. Tree Genet Genomes 9:377–386

Teh HF, Neoh BK, Hong MPL, Low JYS, Ng TLM, Ithnin N, Thang YM, Mohamed M, Chew FT, Yusof HM (2013) Differential metabolite profiles during fruit development in high-yielding oil palm mesocarp. PLoS ONE 8:e61344

Teh HF, Neoh BK, Wong YC, Kwong QB, Ooi TEK, Ng TLM, Tiong SH, Low JYS, Danial AD, Ersad MA, Kulaveerasingam H, Appleton DR (2014) Hormones, polyamines, and cell wall metabolism during oil palm fruit mesocarp development and ripening. J Agric Food Chem 62(32):8143–8152

Ting NC, Jansen J, Mayes S, Massawe F, Sambanthamurthi R, Ooi LCL, Chin CW, Arulandoo X, Seng TY, Alwee SS, Ithnin M, Singh R (2014) High density SNP and SSR-based genetic maps of two independent oil palm hybrids. BMC Genomics 15 (309):1–11

Tohge T, Fernie AR (2015) Metabolomics-Inspired Insight into Developmental, Environmental and Genetic Aspects of Tomato Fruit Chemical Composition and Quality. Plant Cell Physiol 56(9):1681–1696

Tranbarger TJ, Dusser S, Joet T, Argout X, Summo M, Champion A, Cros D, Omore A, Nouy B, Morcillo F (2011) Regulatory mechanisms underlying oil palm fruit mesocarp maturation, ripening, and functional specialization in lipid and carotenoid metabolism. Plant Physiol 156(2):564–584

Turner PD, Gillbanks RA (2003) Field diseases and disorders of oil palm. In: Oil palm cultivation and management, vol 10. The Incorporated Society of Planters, pp 625–727

Vranova E, Hirsch-Hoffmann M, Gruissem W (2011) AtIPD: a curated database of *Arabidopsis* isoprenoid pathway models and genes for isoprenoid network analysis. Plant Physiol 156:1655–1656

Weckwerth W (2011) Green systems biology—rom single genomes, proteomes and metabolomes to ecosystems research and biotechnology. J Proteomics 75:284–305

Wienkoop S, Morgenthal K, Wolschin F, Scholz M, Selbig J, Weckwerth W (2008) Integration of metabolomic and proteomicphenotypes. Mol Cell Proteomics 7(9):1725–1736

Woittiez LS, Van Wij MT, Slingerland M, Van Noordwijk M, Giller KE (2017) Yield gaps in oil palm: a quantitative review of contributing factors. Eur J Agron 83:57–77

Wong YC, Teh HF, Mebus K, Ooi TEK, Kwong QB, Koo KL, Ong CK, Mayes S, Chew FT, Appleton DR, Kulaveerasingam H (2017) Differential gene expression at different stages of mesocarp development in high- and low-yielding oil palm. BMC Genomics 18:1–13

Point Projects (EPP) 2: Increase the national FFB yield. National Key Economic Areas (NKEA). Ministry of Plantation Industries and Commodities, pp 9–10

Zhou W, Chen T, Chong Z, Rohrdanz MA, Melott JM, Wakefield C, Zeng J, Weinstein JN, Meric-Bernstam F, Mills GB, Chen K (2015) TransVar: a multilevel variant annotator for precision genomics. Nat Methods 12(11):1002–1003

The manufacturer's authorised representative in the EU is Springer
Nature Customer Service Centre GmbH, Europaplatz 3, 69115 Heidelberg,
Germany. If you have any concerns regarding our products, please
contact ProductSafety@springernature.com

Printed and bound by CPI Group (UK) Ltd, Croydon, CR0 4YY
30/09/2026
02225665-0018